STABILITY UNDER SEISMIC LOADING

Proceedings of a session at
Structures Congress '86 sponsored by the
Structural Division of the
American Society of Civil Engineers

Hyatt Regency Hotel
New Orleans, Louisiana
September 15-18, 1986

Edited by Franklin Y. Cheng

Published by the
American Society of Civil Engineers
345 East 47th Street
New York, NY 10017-2398

ENGINEERING

ABSTRACT

This publication includes five papers that cover the p-Δ effect on seismic structures with semi-rigid connections, the effect of panel zone deformations on the structural response behavior, assessment of UBC for K-braced structures without backup ductile moment-resisting frames, experimental studies of a structure subjected to multicomponent base excitation, and the correlation between experimental and analytical studies of various structures subjected to simulated earthquakes. Also included in the papers are design criteria and recommendations for future work.

Library of Congress Cataloging-in-Publication Data

Structures Congress '86 (1986 : New Orleans, La.)
    Stability under seismic loading.

    "Hyatt Regency Hotel, New Orleans, Louisiana, September 15-18, 1986."
    Includes index.
    1. Buildings—Earthquake effects—Congresses.
I. Cheng, Franklin Y. II. American Society of Civil Engineers. Structural Division. III. Title.
TA654.6.S87   1986   624.1'762   86-22130
ISBN 0-87262-556-7

FOREWORD

Since its founding in 1944, the principal objectives of the Structural Stability Research Council (SSRC), formerly the Column Research Council, have been to foster research and to assist development of improved design procedures for compressive components of metal structures. The Council provides guidance to practicing engineers and specification writers with refined design procedures as published in the Guide to Stability Design Criteria for Metal Structures. To meet the needs of the research and practice in earthquake structural engineering, Task Group 24 on Stability Under Seismic Loading was established in 1983 and currently consists of the following technical members:

F. Y. Cheng (Chairman)	S. C. Goel	K. Takanashi
W. F. Chen	R. Husid	K. Z. Truman
L. D. Carpenter	L. W. Lu	M. Wakabayashi
S. A. Freeman	S. A. Mahin	

The scope of Task Group is to investigate stability under seismic loading of structures and their component frames and members.

Several papers in the areas of analysis, code assessment, and experimental and field studies were published in SSRC annual technical session proceedings. This ASCE Structures Congress session is a part of the Task Group's activities to introduce relevant problems to the structural engineering community. Each of the papers included in the proceedings has been accepted for publication by the Proceedings Editor. All papers are eligible for discussion in the Journal of Structural Engineering and for ASCE Awards.

Jerry S. B. Iffland of the Executive Committee of both SSRC and ASCE STD was most helpful in making arrangements for this session. The assistance of Shiela Menaker of ASCE Staff and that of the SSRC Task Group members is gratefully acknowledged. The cover photo was taken in Mexico City by L. W. Lu when he and F. Y. Cheng embarked on field studies after the 1985 Mexico earthquakes.

Franklin Y. Cheng
Editor and Session Chairman

CONTENTS

Connection Panel Zone Deformation in Steel Frames

Eric M. Lui [1], A.M.ASCE and Wai-Fah Chen [2], M.ASCE

Abstract

Conventional analyses of frameworks are usually carried out without considering the effect of panel zone deformation on frame behavior. As a result, center-to-center distances rather than clear spans are used for the lengths of the members. As evident from experimental studies, the effect of panel zone deformation has a pronounced influence on frame behavior. In particular, the strength and drift of the frame will be affected if panel zone deformation is taken into consideration in the analysis. In this paper, various deformation modes of the panel zone are identified. A simple finite element model is used to represent all these modes. The validity of this model is established by the comparison of theoretical predictions with experiments on joint subassemblages. Finally, a two-bar frame with different behavioral joint models is analyzed numerically to demonstrate the importance of using realistic joint models in a frame analysis.

Introduction

In a steel frame, if the beam is framed into the flange of the column, there exists a region called the panel zone which is composed of the web and flanges of the column (Fig. 1). The behavior of this panel zone has a significant influence on the behavior of the frame. Figure 2 shows a possible system of forces that acts on the joint panel of an interior beam-to-column connection. Under the action of these forces, the joint panel will deform. The various deformation modes are shown in Fig. 3. In addition to causing deformation, these forces may cause premature yielding of the panel zone resulting in a reduction in strength and stiffness of the frame.

Numerous tests (Becker, 1975, Bertero et al, 1972, Fielding and Huang 1971, Krawinkler, 1978) have been performed in the past decade to investigate the load-deformation behavior of the joint panel using connection subassemblages. Particular attention was given to the shear capacity of the panel zone and the effect ofpanel zone shear deformation on the strength and stiffness of the subassemblages. The significant features observed in these tests are:

---

1 Assistant Professor, Department of Civil Engineering, Syracuse University, Syracuse, New York 13244

2 Professor and Head of Structural Engineering, School of Civil Engineering, Purdue University, West Lafayette, Indiana 47907

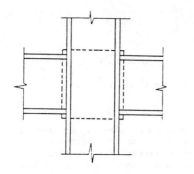

Fig. 1   Panel Zone

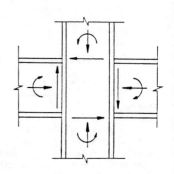

Fig. 2   Forces Acting on a Panel Zone

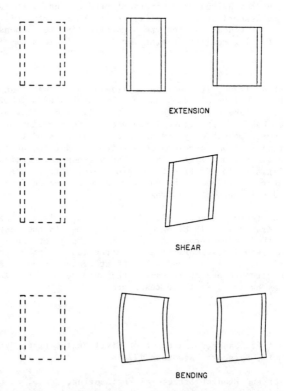

EXTENSION

SHEAR

BENDING

Fig. 3   Deformation Modes of a Joint Panel

1. There are two distinct stiffnesses in the joint shear force-deformation response of the panel. An elastic stiffness, followed by a smaller, almost constant stiffness for a long range of deformation (Fig. 4);

2. Large ductility of the joint panel is observed before failure;

3. Failure is usually caused by fracture of the welds or beam flange on the face of the column flange.

The existence of a second or post yield stiffness in the shear force-deformation response is attributed to the following (Fielding and Chen, 1973, Kato 1982):

1. The resistance of the boundary elements such as the column flanges and stiffeners of the joint panel;

2. The onset of strain-hardening of the web of the joint panel before complete yielding of the boundary elements;

3. The restraint from the adjoining beams and columns.

Based on these observations, a finite element model of the panel zone has been proposed (Lui and Chen, 1986). This model is capable of representing the various modes of deformation depicted in Fig. 3. In addition, yielding and strain-hardening of the web panel are also considered in the present frame analysis with panel zone deformation. The validity of this model will be demonstrated in this paper by comparison with experiments on joint subassemblages. A comprehensive study on the effects of connection flexibility and panel zone deformation on the behavior of plane steel frames is given elsewhere (Lui, 1985).

Basic Assumptions

The assumptions used for the panel zone model are:

1. An elastic-perfectly plastic-strain hardening stress-strain behavior of the web panel is assumed (Fig. 5);

2. Although large rigid body rotation of the joint panel is allowed, the deformation or distortion of the joint panel remains small;

3. No local buckling or lateral torsional buckling of the panel is allowed. In other words, only strength limit state will be considered for the joint panel in the model;

4. Yielding of the web of the joint panel will occur as the state of stress reaches the yield surface described by he von Mises or J_2 theory;

5. Isotropic hardening rule is used to describe the subsequent yield or loading surfaces;

6. Fracture of the material is not considered.

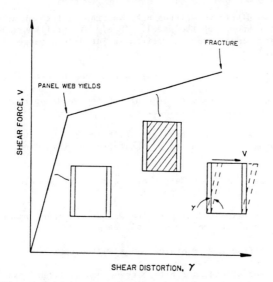

Fig. 4   Typical Shear Force-Distortion Behavior of a Joint Panel

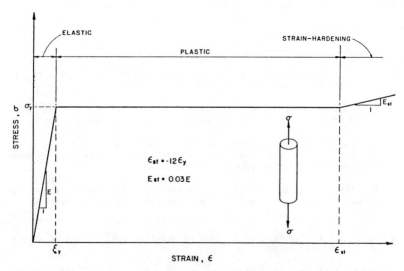

Fig. 5   Assumed Uniaxial Stress-Strain Behavior

The details of the finite element formulation of the panel zone deformation model are given elsewhere (Lui and Chen, 1986) and will therefore not be repeated here.

Numerical Studies

     The panel element developed by Lui and Chen (1986) is used here in
conjunction with the frame and connection elements developed recently
by Lui (1985) to investigate the behavior of frames. A load-controlled
incremental Newton-Raphson iterative solution technique is employed in
the numerical analysis. Detailed discussion of the solution algorithm
is given elsewhere (Lui, 1985). Herein, the validity of the panel zone
model is first verified by comparison of the numerical results with
experiments. A simple two-bar frame is then analyzed using different
behavioral models regarding the deformational behavior of the connection and panel zone to demonstrate the effect of joint flexibility on
the behavior of the frame.

Comparison with Experiments

(i) Lehigh Test

     Shown in the inset of Fig. 6 is a subassemblage used by Fielding
and Huang (1971) to investigate the behavior of the joint panel. The
joint details of the subassemblage is given in the reference cited.
The column was first loaded with a compressive force of 0.5 P_y where P_y
is the yield load of the column. The beam load was then applied at the
free end of the beam until failure. During the entire beam loading
phase, the column load was maintained at $0.5P_y$. As a result, the panel
zone is under a combined loading of axial force from the column load
and shear and moment from the beam load. The experimental load-
deflection behavior of this subassemblage is shown as solid line in
Fig. 6. Yielding of the web of the panel zone occurs at a load of 86
kips (383 kN) after which a definite decrease in stiffness of the joint
panel was observed. At 150 kips (667 kN), cracks were observed at the
ends of the top horizontal stiffeners and the specimen was unloaded.
Also shown in this figure as dashed line is the numerically obtained
load-deflection curve. It can be seen that good correlation exists
between the results obtained numerically and experimentally both in the
elastic and post-yield regimes. In the numerical solution, the actual
measured material properties were used. The apparent larger stiffness
and higher yield load obtained numerically as compared to test is
attributed to the assumption of full fixity at both ends of the column
in the numerical model, whereas in reality, noticeably column end rotations were observed (Fielding and Huang, 1971).

(ii) Berkeley Tests

     The insets of Figs. 7 and 8 show two subassemblages (Specimens A
and B) of an experimental investigation of the behavior of panel zone
by Bertero et al. (1972). Specimen A is typical of an upper story and
Specimen B is typical of a lower story. For Specimen A, an axial force
of $0.36P_y$, where P_y is the yield load of the column, was applied to the
column and vertical beam forces of approximately 6 kips (27 kN) were

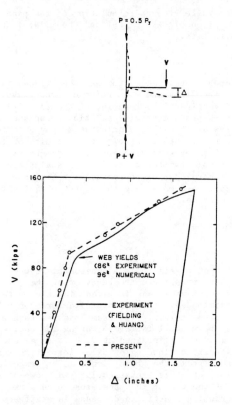

Fig. 6   Comparison of Numerical and Experimental Connection
Subassemblage Load-Deflection Behavior
(Fielding and Huang)   (1 kip = 4.45 kN, 1 in = 25.4 mm)

applied at every third-point on the beams. For Specimen B, an axial force of 0.48P$_y$ was applied to the column and vertical beam forces of approximately 6 kips (27 kN) were applied at every third-point on the beams. These subassemblages were then subjected to a horizontal force H applied cyclically at the free end of the column. Since the present analysis deals only with monotonic loadings, only the initial branch of the hysteresis load-deflection behavior of these subassemblages will be investigated. The experimental curves are shown as solid lines in Figs. 7 and 8 for Specimen A and B, respectively. For Specimen A (Fig. 7), extensive yielding of the web of the joint panel was observed in the test. The numerically obtained load-deflection curve using the actual measured material properties is shown as dashed line in the figure. Good agreement between the two curves is observed. The numerically predicted yielding of the web occurs at H = 6.4 kips (28 kN) which agrees favorably with the experiment. Since no stiffeners are used in the panel zone, the structure becomes numerically unstable as soon as the panel yields (that is, as soon as a panel hinge is formed). Consequently, the numerically generated load-deflection curve shows a plateau at H = 6.4 kips (28 kN).

The numerical and experimental load-deflection curves for Specimen B (Fig. 8) also agree well with each other. Failure of this specimen was due to the formation of plastic hinges in the beams. In the numerical solution, convergence of solution became impossible when a local collapse mechanism developed in the left beam at H = 21.2 kips (94.3 kN). However, in the experiment, additional load can be applied because even after the formation of a local collapse mechanism in the left beam, the subassemblage is still statically determinate and so total collapse will not occur until a third hinge formed in the right beam. Consequently, the experimental load-deflection curve rises above H = 21.2 kips (94.3 kN) in Fig. 8.

Although the post-yield behavior of the specimen shown in Fig. 8 cannot be predicted by the model due to local failure of the beam. The model does give a good representation of the elastic behavior of the subassemblage.

In practical design, it is customary to investigate the drift (that is, lateral displacement) of a frame under service loading conditions to ensure that the frame will not deflect excessively so as to cause discomfort to the occupants or overstress in the connecting elements. In the following section, a simple two-bar frame will be analyzed using a number of different behavioral models to demonstrate the importance of panel zone deformation on the drift of a structure.

Analysis of a Two-Bar Frame

Shown in Fig. 9 is a series of structural models for a two-bar frame. Model 1 (M1) is the conventional structural model commonly used by engineers and designers. The connection joining the beam and column is assumed to be rigid and center-to-center distances are used for the lengths of the beam and column. Model 2 (M2) is a more refined model. Although center-to-center distances are still used for the lengths of the beam and column, the connection is modeled by a spring having a

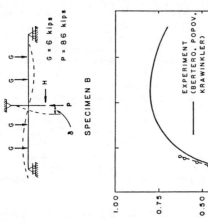

Fig. 7 Comparison of Numerical and Experimental Connection Subassemblage Load-Deflection Behavior (Specimen B, Bertero, et al.) (1 kip = 4.45 kN, 1 in = 25.4 mm)

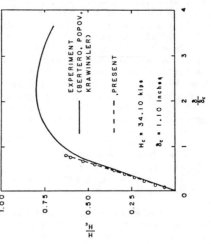

Fig. 8 Comparison of Numerical and Experimental Connection Subassemblage Load-Deflection Behavior (Specimen B, Bertero, et al.) (1 kip = 4.45 kN, 1 in = 25.4 mm)

rotational stiffness as described by the nonlinear moment-rotation behavior of the connection shown in Fig. 10. Model 3 (M3) assumes that the connection is rigid but the finite size and deformable capability of the panel zone are taken into account. In this model, clear spans are used for the lengths of the beam and column. Model 4 (M4) is the most refined model of all. The connection is modeled as flexible with rotational stiffness described by the nonlinear moment-rotational behavior of the connection and the joint panel is modeled as deformable with clear spans used for the lengths of the beam and column.

For Models 2 and 4, the moment-rotation characteristics of four commonly used connections (double web angles, top and seat angles, end plate and T-stub) compatible with the beam and column sections (W14 x 34) used are shown in Fig. 10. They are labeled as C1, C2, C3 and C4, respectively. C1 represents a very flexible and C4 represents a very stiff connection.

The two-bar frame is loaded by a horizontal force H as shown in Fig. 9, the load-deflection behavior of the frame using different models and connections are shown in Figs. 11 to 14. Failure of the structure is due to the formation of a plastic hinge at the junction of the beam and column. For Connections C1 and C2, the plastic hinge developed in the connection when the ultimate capacity of the connection was exhausted. However, for Connections C3 and C4, the plastic hinge developed in the column at the beam and column juncture because the ultimate moment capacities of these two connections exceed the plastic moment capacities of the sections. This explains why the ultimate loads of the subassemblages using Connections C1 and C2 are less than that of the subassemblages using Connections C3 and C4. For C1 and C2, the ultimate load is dictated by the strength of the connections, whereas for C3 and C4, the ultimate load is controlled by the strength of the sections.

If we examine Figs. 11 to 14 carefully, we can conclude that Model 2 will give a satisfactory approximation to Model 4 provided that the connections are relatively flexible when compared to the adjoining beam and column (e.g., C1 and C2). This is because the contribution to additional drift from the flexible connection far outweighs the contribution from panel zone deformation. However, for relatively stiff connections (e.g., C3 and C4), Model 2 and Model 3 will give comparable results which means that the contribution to additional drift from panel zone distortion will be as important as that from connection flexibility. As a rough estimate, if $(R_{ki} L/EI) \leq 5$, where R_{ki} is the initial stiffness of the connection and L, E, I are respectively the length, elastic modulus and moment of inertia of the member to which the connection is attached, then Model 2 is justified to be used in place of Model 4.

To examine the effect of panel zone deformation on the drift of the frame, the load-deflection behavior of the two-bar frame assuming rigid connection using Model 1 and Model 3 are shown in Fig. 15. It can be seen that at working load, 13% drift of the frame is due to panel zone deformation. This indicates that panel zone deformation should be considered in the design of moment-resisting frames.

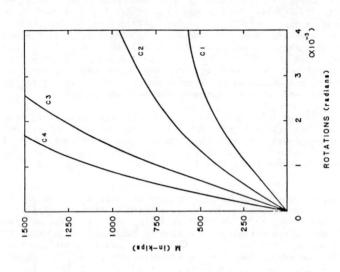

Fig. 10   Connection Moment-Rotation Behavior
          Used for the Two-Bar Frame
          (1 in-kip = 113 N-m)

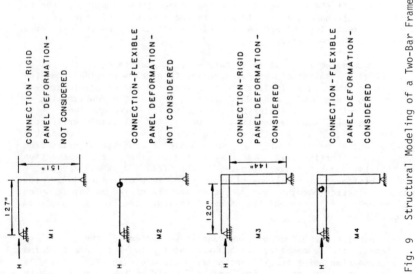

Fig. 9   Structural Modeling of a Two-Bar Frame
         (1 in = 25.4 mm)

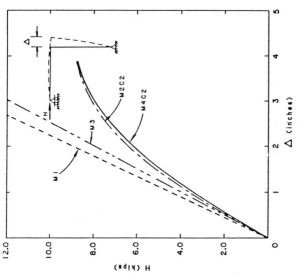

Fig. 12 Load-Deflection Behavior of the Two-Bar Frame Using Connection C2 (1 kip = 4.45 kN, 1 in = 25.4 mm)

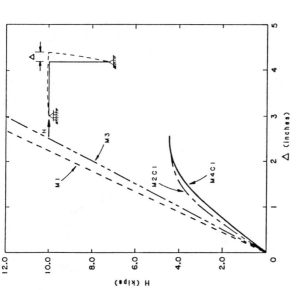

Fig. 11 Load-Deflection Behavior of the Two-Bar Frame Using Connection C1 (1 kip = 4.45 kN, 1 in = 25.4 mm)

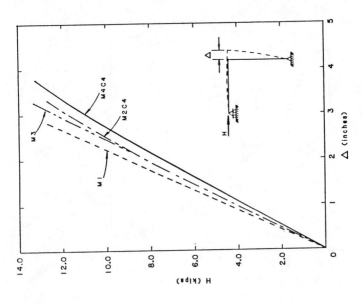

Fig. 14   Load-Deflection Behavior of the Two-Bar
          Frame Using Connection C4
          (1 kip = 4.45 kN, 1 in = 25.4 mm)

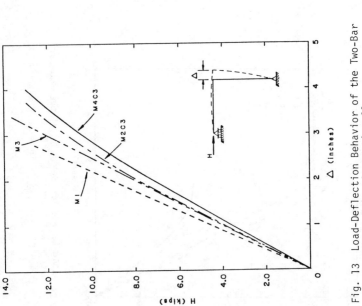

Fig. 13   Load-Deflection Behavior of the Two-Bar
          Frame Using Connection C3
          (1 kip = 4.45 kN, 1 in = 25.4 mm)

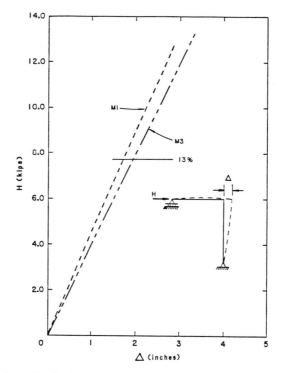

Fig. 15 Effect of Panel Zone Deformation on the Drift of the
Two-Bar Frame (1 kip = 4.45 kN, 1 in = 25.4 mm)

Summary and Conclusions

The behavior of the panel zone is first described and an analytical model that has been developed recently to represent the panel zone deformational behavior (extensional, shear, bending) is utilized here in a frame analysis. The validity of the model is shown by comparing the subassemblage frame analysis with experiments and the importance of the panel zone deformation on the drift of a structure under service load is demonstrated by the analysis of a two-bar frame.

Experiments on connection subassemblages conducted on the past decade have shown that the panel zone plays an important role in affecting the serviceability and ultimate behavior of moment-resisting frames. Because of the deformational and inelastic behavior of the panel zone, the drift and strength of the frame will be affected. The effects of panel zone deformation and inelasticity are usually undesirable since they will cause an increase in frame drift and a decrease in frame strength. In this paper, a simple finite element model that represents accurately the behavior of the panel zone was implemented in a structural analysis program. The inclusion of the panel zone deformation in a frame analysis increases the number of degrees of freedom of the analytical model, but it does provide a much better prediction of the frame behavior.

For moment-resisting frames in which connections of high rigidities are used, panel zone deformation is an important factor to be considered in the analysis and design of the frames.

Appendix I - References

1. Becker, R., "Panel Zone Effect on the Strength and Stiffness of Steel Rigid Frames", Engineering Journal, AISC, Vol. 12, No. 1, First Quarter, 1975, pp. 19-29.

2. Bertero, V.V., Popov, E.P., and Krawinkler, H., "Beam-Column Subassemblages Under Repeated Loading", Journal of the Structural Division, Proc. ASCE, Vol. 98, No. ST5, May 1972, pp. 1137-1159.

3. Fielding, D.J. and Huang, J.S., "Shear on Beam-to-Column Connections", The Welding Journal, Fol. 50, July 1971.

4. Fielding, D.J. and Chen, W.F., "Steel Frame Analysis and Connection Shear Deformation", Journal of the Structural Division, ASCE, Vol. 99, No. ST1, January 1973, pp. 1-18.

5. Kato, B., "Beam-to-Column Connection Research in Japan", Journal of the Structural Division, ASCE, Vol. 108, No. ST2, February 1982, pp. 343-360.

6. Krawinkler, H., "Shear in Beam-Column Joints in Seismic Design of Steel Frames", Engineering Journal, AISC, Vol. 15, No. 3, Third Quarter, 1978, pp. 82-91.

7. Lui, E.M., "Effects of Connection Flexibility and Panel Zone Deformation on the Behavior of Plane Steel Frames", Thesis presented to Purdue University at West Lafayette, IN, in partial fulfillment of the requirements for the Degree of Doctor of Philosophy, 1985.

8. Lui, E.M. and Chen, W.F., "Frame Analysis with Panel Zone Deformation", International Journal of Solids and Structures", Vol. 22, 1986.

SEISMIC DESIGN OF BRACED STEEL FRAMES

Subhash C. Goel[*], M. ASCE, and Xiaodong Tang[**]

ABSTRACT

Seismic behavior of a six-story, K-braced structure without backup ductile moment-resisting frames designed according to the current Uniform Building Code is studied in this paper. The associated problems of column buckling and fracture of bracing members during a severe earthquake motion are discussed. A practical and rational method for safe design of such structures is presented.

Introduction

According to the current Uniform Building Code [6] braced steel buildings upto 160 feet in height in Seismic Zone Nos. 3 and 4 can be designed in one of two ways. Dual systems consisting of ductile moment-resisting frames and braced frames are designed with the horizontal force factor equal to 0.8. Alternatively, braced frames without the backup ductile moment frames can be designed with the horizontal force factor equal to 1.0. These latter type of structures are quite common and inspite of increased design seismic force they often turn out to be more economical because of simple connections between the beams and columns. Inspite of their popularity not many studies have been performed in the past regarding their performance during severe earthquakes. This paper makes a brief attempt in this direction.

Frame F1

A six-story, full-scale test structure was used in the U.S.-Japan Cooperative Earthquake Research Program. It was designed to act as a dual system in the direction of loading. It consisted of three ductile moment resisting frames with concentric K-bracing in one bay of the middle frame in Phase 1 (shown in Fig. 1) and eccentric bracing in the other bay in Phase 2. The structure was designed to satisfy the requirements of both the 1979 Uniform Building Code and the 1981 Japanese building code [2].

---

 * Professor of Civil Engineering, The University of
   Michigan, Ann Arbor, MI.
** Doctoral Candidate, Department of Civil Engineering,
   The University of Michigan, Ann Arbor, MI.

The structure selected for this study is patterned after the U.S.-Japan test structure with concentric K-bracing except that it has no backup moment-resisting frames. The floor plan, dimensions and gravity loads are kept the same. The structure is first designed according to the requirements of current Uniform Building Code, 1982 edition [6], and allowable design procedure of the current AISC Specification [1]. W sections in A36 steel are used for beams and columns, and square structural tubes of A500 grade B steel for the bracing members. The connections of beams with columns are simple, non-moment type. The resulting member sizes of the braced frame are shown in Fig. 2. This frame which provides all the lateral force resistance for the entire structure is designated as F1 in this study.

The inelastic response of Frame F1 is computed for the NS component of the 1978 Miyagi-ken Oki earthquake with the peak acceleration scaled to 500 gals (1 gal = 1 cm/sq. cm). This ground motion was also used in the final pseudodynamic test of the U.S.-Japan concentrically braced structure. Some response results of Frame F1 are shown in Figs. 3 and 4. DRAIN-2D program originally developed at UC Berkeley [8] and later enhanced at the University of Michigan [3,7] was used for this study. It is noticed from Fig. 3 that the horizontal displacements at upper three floors are quite large. Few plastic hinges formed in the beams and inspite of large displacements at the upper floors brace buckling occurred only in the second and third stories, Fig. 4(a). This figure also shows that plastic hinges formed at a number of locations in the columns even though the columns received only small moments from the bracing members. This indicates that columns in this frame were subjected to large axial forces which led to buckling in three of them, also shown in Fig. 4(a). Column buckling commenced as early as 4 seconds into the response. Buckling in columns resulted in large tilt of the frame in upper four stories as shown in Fig. 4(a).

Buckling of columns is perhaps the most significant and undesirable aspect of the response of Frame F1 resulting in large displacements at upper floor levels. Buckling in columns (local or overall) due to large axial force reversals may cause instability and complete collapse of braced structures under certain conditions. This aspect seems to have played quite important role in the observed behavior of Pino Suarez buildings during the 1985 Mexico City earthquake.

Frame F2

In order to eliminate the problem of column buckling during a severe earthquake motion the design procedure for columns was modified in Frame F2. In this frame the braces and beams were kept the same as in Frame F1. An upper bound on the required ultimate strength for the columns is calculated due to 1.3 times the design dead and live loads, and the vertical component of maximum compressive strength of the braces. The sections for columns in the braced bay were then selected according to the requirements of AISC Specification Part 2 (Plastic Design). This frame is called F2 and the member sizes are shown in Fig. 5. Increase in column sizes in the braced bay as compared with those in Frame F1 is apparent.

Response of Frame F2 to the same ground motion is shown in Figs. 6 and 7. It is noticed that inspite of formation of plastic hinges in beams as well as columns no buckling occurred in the columns. However, the braces underwent cyclic buckling in all stories except the top one. Horizontal floor displacements are significantly smaller than those in F1. Axial deformation history of the two braces in second story is shown in Fig. 8. Recent studies have shown that severe local buckling and fractures in tubular bracing members of similar proportions occur in the first few cycles of post-buckling deformation [4,5]. Thus, in "real life" the bracing members of Frame F2 will undergo early fractures which would lead to complete collapse since the lateral stability of this structure solely depends on the integrity of its bracing members.

Frame F3

Recent tests on tubular bracing members at the University of Michigan and briefly reported by Goel [5] show that the ductility and energy dissipation capacity of these members can be much improved by delaying local buckling. This can be achieved by using significantly smaller width-thickness ratios or by filling the tubes with a stiff material such as concrete. Thus, the tubular bracing members of Frame F3 were designed by limiting the width-thickness ratios to $95/\sqrt{F_y}$, which is exactly half of that allowed by the current AISC Specification Part 2 [1]. For nominal yield strength of 46 ksi for A500 grade B steel this would amount to a limit of 14.

Modern building codes generally allow smaller design seismic forces for ductile structures while imposing "penalties" for less ductile structures or structural elements. According to the current Uniform Building Code buildings having ductile moment resisting space frames can be designed with a horizontal force factor of 0.67 or 0.8. For buildings in Seismic Zone Nos. 3 and 4, and for buildings with importance factor greater than 1.0 in Zone No. 2, all members in braced frames must be designed for 1.25 times the force determined otherwise. Since, the ductility of the bracing members in Frame F3 is ensured by using lower width-thickness ratios it was decided to delete the "penalty" factor of 1.25 in their design. The columns were designed by following the procedure as in Frame F2. The resulting member sizes for the braced frame F3 are shown in Fig. 9.

The computed response of Frame F3 is shown in Figs. 10-12. It is noticed that the horizontal floor displacements are about the same as in Frame F2. Plastic hinges formed in several beams and columns but no column buckling was encountered. Cyclic buckling occurred in all braces. Axial deformation history of the two braces in second story is shown in Fig. 12, which is almost the same as in Frame F2. However, due to much more compact sections their integrity is ensured. Thus, it can be concluded that Frame F3 will perform satisfactorily when subjected to a severe ground motion such as the one used in this study.

Conclusions

Based on the results and discussion presented in this paper the following conclusions can be drawn:

1. Buildings consisting of braced frames without backup ductile moment resisting frames may not survive a severe earthquake because of column buckling and/or early failure of bracing members.

2. Column buckling can be prevented by using a limit state design procedure such as the one suggested in this paper.

3. If the ductility of bracing members is ensured their design can be based on forces smaller than those specified by current building codes.

Acknowledgement

This study was sponsored by the National Science Foundation under Grant No. ECE-8516866. The conclusions and opinions expressed in this paper are solely those of the authors and do not necessarily represent the views of the sponsor.

References

1. AISC, "Specification for the Design Fabrication and Erection of Structural Steel for Buildings," American Institute of Steel Construction, Chicago, Ill., 1978.

2. Askar, G., Lee, S.J., and Lu, Le-Wu, "Design Studies of the Six-Story Steel Test Building," Report No. 467.3, Fritz Engineering Laboratory, Lehigh University, Bethlehem, Pennsylvania, June 1983.

3. Boutros, M.K., and Goel, S.C., "Analytical Modelling of Braced Steel Structures," Report No. UMCE 85-7, Department of Civil Engineering, University of Michigan, Ann Arbor, Mich., August 1985.

4.   Foutch, D.A., Yamanouchi, H., Roeder, C.W., Midorikawa,
     M., Nishiyama, I., and Watabe, M., "Summary of Design
     and Construction of the Full-Scale Specimen and Results
     of the Phase 1 Tests," ASCE Structures Congress III,
     San Francisco, Calif., October, 1-5, 1984.

5.   Goel, S.C., "Seismic Stability of Braced Steel
     Structures," Proceedings, Structural Stability Research
     Council Meeting, Washington, D.C., April 15-16, 1986.

6.   International Conference of Building Officials, "Uniform
     Building Code," Whittier, Calif., 1982.

7.   Jain, A.K., and Goel, S.C., "Hysteresis Models for Steel
     Members subjected to Cyclic Buckling or Cyclic End
     Moments and Buckling," Report No. UMEE 78R6, Department
     of Civil Engineering, University of Michigan, Ann Arbor,
     Mich., December 1978.

8.   Kanaan, A.E., and Powell, G.H., "General Purpose
     Computer Program for Inelastic Dynamic Response of Plane
     Structures," Report No. EERC 73-6, University of
     California, Berkeley, Calif., April 1973.

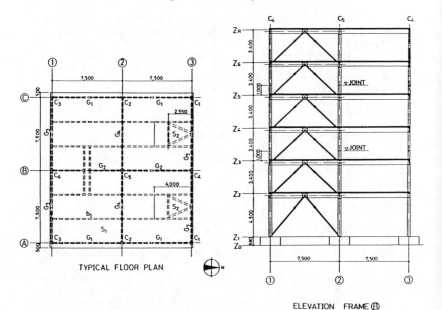

Fig. 1   U.S.-Japan Concentric Braced Structure.

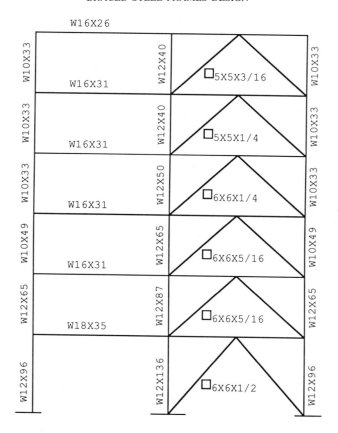

Fig. 2   Member Sizes, Frame F1

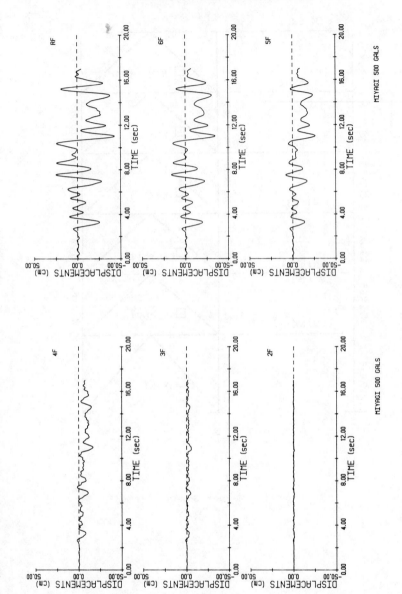

Fig. 3   Horizontal Floor Displacements, Frame F1

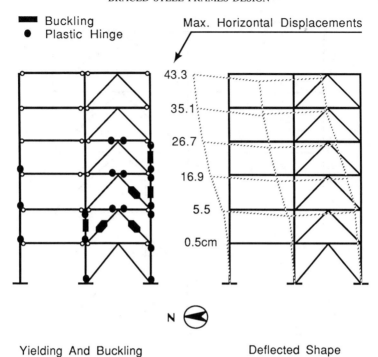

Fig. 4   Response Of Frame F1 to Miyagi-Ken-Oki
           Earthquake (t=11 sec)

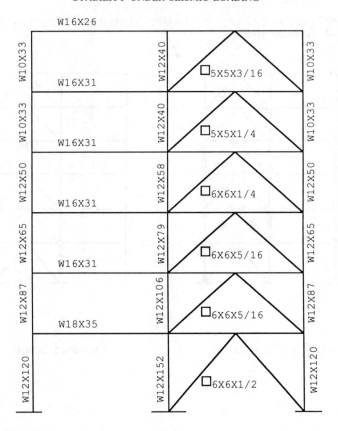

Fig. 5   Member Sizes, Frame F2

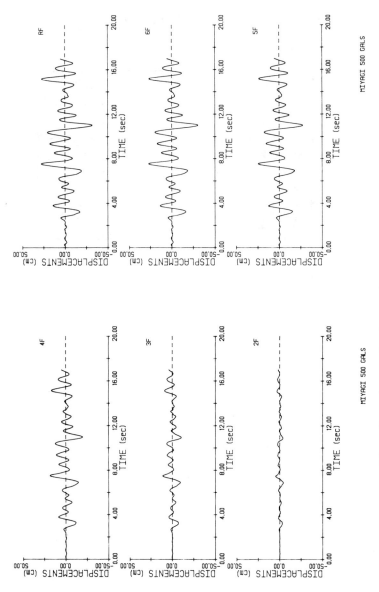

Fig. 6 Horizontal Floor Displacements, Frame F2

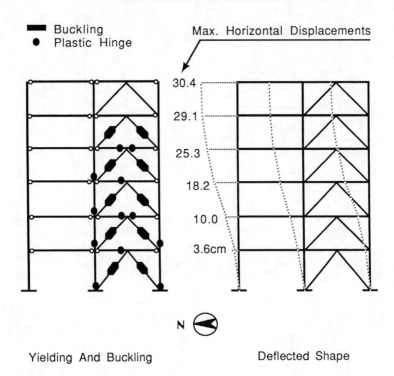

Fig. 7   Response Of Frame F2 to Miyagi-Ken-Oki
           Earthquake (t=11 sec)

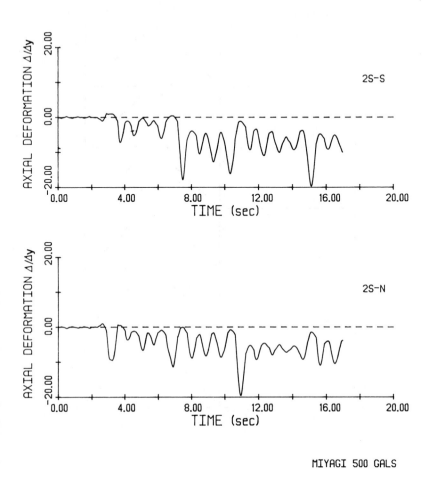

MIYAGI 500 GALS

Fig. 8 Brace Deformation History, Second Story, Frame F2

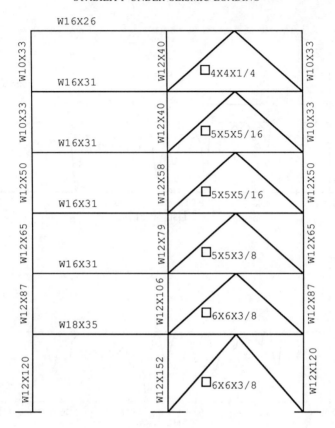

Fig. 9   Member Sizes, Frame F3

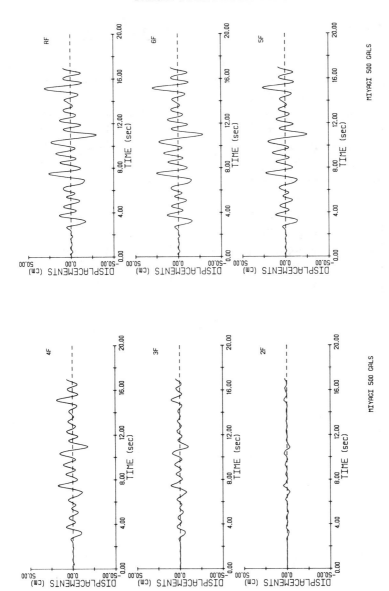

Fig. 10   Horizontal Floor Displacements, Frame F3

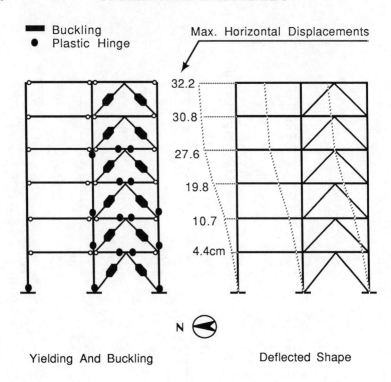

Fig. 11   Response Of Frame F3 to Miyagi-Ken-Oki
           Earthquake (t=11 sec)

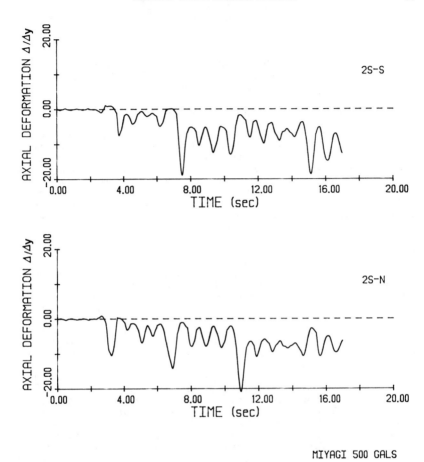

MIYAGI 500 GALS

Fig. 12   Brace Deformation History, Second Story, Frame F3.

THE EFFECTS OF P-DELTA AND SEMI-RIGID CONNECTIONS
ON RESPONSE BEHAVIOR OF INELASTIC STEEL FRAMES
SUBJECTED TO CYCLIC AND SEISMIC LOADINGS

By F.Y. Cheng,[1] M. ASCE, and D.S. Juang,[2] A.M. ASCE

ABSTRACT: A numerical method is presented for analyzing semi-rigid
connected steel frames subjected to various time-dependent excitations
with consideration of damping, nonlinear material behavior for members
and connections, and geometric nonlinearity. The material nonlinearity
is modeled as bilinear. The external excitations may be cyclic loads or
seismic motions. The structural formulation is based on the displacement
method and the lumped mass model. Numerical examples are provided that
demonstrate the significant effects of various considerations on the
dynamic response.

INTRODUCTION

In seismic steel frame design for strong-columns and weak-girders,
the inelastic deformations are usually assumed to be developed at the
girders and at the girder connections such that the energy can be
dissipated through hysteretic action in the girders and the connections.
The connections can be constructed as fully penetrated welds or bolts.
Bolted connections are essential in many locations because climatic
conditions may limit year-round field weldings and field-bolted frames
may be preferred. Bolted joints are flexible and are considered to be
semi-rigid connections. In the past, publications mainly emphasized on
experimental behavior of various types of semi-rigid connections and the
analytical models were relatively simple; they included neither the nonlinear behavior of members nor the large deflection due to P-Δ effects
(6-11). This paper developes the dynamic analysis of semi-rigid connected frames with consideration of large deflections and nonlinear
stiffnesses in both joints and members. The displacement method and
the lumped mass model are employed in the structural formulation for a
general approach of various types of frameworks subjected to static and
dynamic forces.

STRUCTURAL FORMULATION

1. Stiffness Coefficients of Elastic Members with Bilinear Semi-
Rigid Joints and P-Δ Effect -- Based on the experimental studies, the
moment-rotation relationship of a semi-rigid joint may be sketched in
Fig. 1 where the moment, M, transmitted by the connection is in function
of the slip angle, θ', that is defined as the rotation between the

---

1. Professor, Department of Civil Engineering, University of Missouri-
   Rolla, Rolla, Missouri, U.S.A.
2. Associate Professor, Civil Engineering Department, National Central
   University, Taiwan; formerly Graduate Research Assistant at the
   University of Missouri-Rolla.

tangent line of the connected member and the center line of the connection. The curve presented in this figure indicates that the moment-rotation relationship at the beginning is nearly linear, followed by nonlinear variation, and then becomes asymptotic linear strain-hardening envelope (4,8,10). A bilinear model (1,5) can be imposed on this curve to approximate the calculation for which a_1 is the initial slope, a_2 is the second slope, M_t is the transition moment, and M_f is the failure moment.

Let us consider a typical member as shown in Fig. 2, in which θ_i is the joint rotation, θ_i' the slip angle, θ_i'' is the rotation of member end i, and $\theta_i'' = \theta_i - \theta_i'$. The moments transmitted by the connections can be expressed as

$$M_i = \frac{4EI}{\ell} (\theta_i - \theta_i') \tag{1}$$

Let $a_1 = f_i EI/\ell$, where f_i = the constant which describes the stiffness of the connection; then the moment can be expressed as

$$M_i = a_1 \theta_i' = \frac{f_i EI \theta_i'}{\ell} \tag{2}$$

Substituting Eq. (2) into (1), the relationship between slip-angle and joint rotation is expressed as

$$\theta_i' = \frac{4}{4 + f_i} \theta_i \tag{3}$$

Let us define the percentage of rigidity, τ, as the ratio of the moment, M, that is required to produce a unit joint rotation with a semi-rigid connection divided by the moment, M', that is required to produce a unit joint rotation with a fully rigid connection. Because $M_i' = 4EI/\ell$, the percentage of rigidity can be found by using Eqs. (1) and (3) as

$$\tau_i = \frac{M_i}{M_i'} = 1 - \frac{4}{4+f_i} \tag{4}$$

According to Eq. (2), the slip-angles at member ends can be expressed as

$$\begin{Bmatrix} \theta'_i \\ \theta'_j \end{Bmatrix} = \frac{\ell}{EI} \begin{bmatrix} \dfrac{1}{f_i} & 0 \\ 0 & \dfrac{1}{f_j} \end{bmatrix} \begin{Bmatrix} M_i \\ M_j \end{Bmatrix}$$

$$= [F_c]\{FM\} \tag{5}$$

Let $[\theta\ Y]^T = [\theta_i\ \theta_j\ y_i\ y_j]^T$ be the joint deformation vector, then the member end deformations are

$$\begin{Bmatrix} \theta'' \\ Y \end{Bmatrix} = \begin{Bmatrix} \theta \\ Y \end{Bmatrix} - \begin{Bmatrix} \theta' \\ 0 \end{Bmatrix} \tag{6}$$

Substituting Eq. (6) into the force-deformation relationship of an elastic member yields

$$\begin{Bmatrix} FM \\ FV \end{Bmatrix} = [S] \begin{Bmatrix} \theta'' \\ Y \end{Bmatrix} = [S]\ (\begin{Bmatrix} \theta \\ Y \end{Bmatrix} - \begin{Bmatrix} \theta' \\ 0 \end{Bmatrix}) \tag{7}$$

in which $[FM\ FV]^T = [M_i\ M_j\ V_i\ V_j]^T$, and

$$[S] = \frac{EI}{\ell} \begin{bmatrix} 4 & 2 & -\dfrac{6}{\ell} & -\dfrac{6}{\ell} \\ & 4 & -\dfrac{6}{\ell} & -\dfrac{6}{\ell} \\ & & \dfrac{12}{\ell^2} & \dfrac{12}{\ell^2} \\ \text{Sym.} & & & \dfrac{12}{\ell^2} \end{bmatrix}$$

By substituting Eq. (5) into Eq. (7) and rearrange it, one finds

$$\begin{bmatrix} (1+\dfrac{4}{f_i}) & \dfrac{2}{f_j} & 0 & 0 \\ \dfrac{2}{f_i} & (1+\dfrac{4}{f_j}) & 0 & 0 \\ -\dfrac{6}{\ell f_i} & -\dfrac{6}{\ell f_j} & 1 & 0 \\ -\dfrac{6}{\ell f_i} & -\dfrac{6}{\ell f_j} & 0 & 1 \end{bmatrix} \begin{Bmatrix} M_i \\ M_j \\ V_i \\ V_j \end{Bmatrix} = [S] \begin{Bmatrix} \theta \\ Y \end{Bmatrix} \tag{8}$$

Solving for the force vector from the above equation, the force-deformation relationship for a member with semi-rigid connections can be obtained as

$$
\begin{Bmatrix} M_i \\ M_j \\ V_i \\ V_j \end{Bmatrix} = \begin{bmatrix} m_1 & n & r_1 & r_1 \\ n & m_2 & r_2 & r_2 \\ r_1 & r_2 & s & s \\ r_1 & r_2 & s & s \end{bmatrix} \begin{Bmatrix} \theta_i \\ \theta_j \\ y_i \\ y_j \end{Bmatrix} \tag{9a}
$$

or

$$
\begin{Bmatrix} FM \\ FV \end{Bmatrix} = [S_r] \begin{Bmatrix} \theta \\ Y \end{Bmatrix} \tag{9b}
$$

in which

$$
m_1 = 4S_{ii}, \tag{9c}
$$

$$
m_2 = 4S_{jj}, \tag{9d}
$$

$$
n = 2S_{ij}, \tag{9e}
$$

$$
r_1 = -(4S_{ii}+2S_{ij})/\ell, \tag{9f}
$$

$$
r_2 = -(4S_{jj}+2S_{ij})/\ell, \tag{9g}
$$

$$
s = 4(S_{ii}+S_{jj}+S_{ij})/\ell^2, \tag{9h}
$$

$$
S_{ii} = \tau_i(3+\tau_j)/(4-(1-\tau_i)(1-\tau_j)), \tag{9i}
$$

$$
S_{jj} = \tau_j(3+\tau_i)/(4-(1-\tau_i)(1-\tau_j)), \text{ and} \tag{9j}
$$

$$
S_{ij} = S_{ji} = 4\tau_i\tau_j/(4-(1-\tau_i)(1-\tau_j)). \tag{9k}
$$

Equation (9) was developed on the basis of the first slope, a_1. If the transition moment, M_t, is reached, one may change the stiffness matrix by changing a_1 to the second slope, a_2, the stiffness of the connection in the plastic range. In nonlinear dynamic analysis, the state of yielding criteria for the connections is based on the $2M_t$ elastic stress range as shown in Fig. 1. If the failure moment, M_f, is reached, it indicates that the connection has failed.

When the P-Δ effect is considered, the force-deformation relationship of an elastic member is changed to

$$
\begin{Bmatrix} M_i \\ M_j \\ V_i \\ V_j \end{Bmatrix} = \left(\left(\frac{EI}{\ell}\right) \begin{bmatrix} 4 & & & \text{sym.} \\ 2 & 4 & & \\ -\frac{6}{\ell} & -\frac{6}{\ell} & \frac{12}{\ell^2} & \\ -\frac{6}{\ell} & -\frac{6}{\ell} & \frac{12}{\ell^2} & \frac{12}{\ell^2} \end{bmatrix} - N \begin{bmatrix} \frac{2\ell}{15} & & & \text{sym.} \\ -\frac{\ell}{30} & \frac{2\ell}{15} & & \\ -\frac{1}{10} & -\frac{1}{10} & \frac{6}{5\ell} & \\ -\frac{1}{10} & -\frac{1}{10} & \frac{6}{5\ell} & \frac{6}{5\ell} \end{bmatrix} \right) \begin{Bmatrix} \theta''_i \\ \theta''_j \\ y_i \\ y_j \end{Bmatrix}
$$

$$
= ([S] - [S_g]) \begin{Bmatrix} \theta'' \\ Y \end{Bmatrix} \tag{10}
$$

in which $[S_g]$ is the geometric stiffness matrix and N is the axial force acting on the member with compression direction as positive. By using Eq. (10), the stiffness matrix for a member with semi-rigid connections and the P-Δ effect can be derived which is similar to Eq. (9) except

$$
m_1 = \frac{1}{D} \left(\frac{4EI}{\ell} - \frac{2N\ell}{15} + \left(\frac{4EI}{\ell} - \frac{2N\ell}{15} \right)^2 g_j - \left(\frac{2EI}{\ell} + \frac{N\ell}{30} \right)^2 g_j \right), \tag{10a}
$$

$$
m_2 = \frac{1}{D} \left(\frac{4EI}{\ell} - \frac{2N\ell}{15} + \left(\frac{4EI}{\ell} - \frac{2N\ell}{15} \right)^2 g_i - \left(\frac{2EI}{\ell} + \frac{N\ell}{30} \right)^2 g_i \right), \tag{10b}
$$

$$
n = \frac{1}{D} \left(\frac{2EI}{\ell} + \frac{N\ell}{30} \right), \tag{10c}
$$

$$
r_1 = -\frac{m_1 + n}{\ell}, \tag{10d}
$$

$$
r_2 = -\frac{m_2 + n}{\ell}, \tag{10e}
$$

$$
s = -\frac{r_1 + r_2}{\ell} - \frac{N}{\ell}, \tag{10f}
$$

$$
D = \left(1 + \left(\frac{4EI}{\ell} - \frac{2N\ell}{15} \right) g_i \right) \left(1 + \left(\frac{4EI}{\ell} - \frac{2N\ell}{15} \right) g_j \right) - \left(\frac{2EI}{\ell} + \frac{N\ell}{30} \right)^2 g_i g_j, \tag{10g}
$$

$$
g_i = \ell/EIf_i, \text{ and} \tag{10h}
$$

$$
g_j = \ell/EIf_j. \tag{10i}
$$

2. _Stiffness Coefficients of Bilinear Members with Semi-rigid Joints and P-Δ Effect_ -- The material property of structural members is assumed to be bilinear as used in Cheng's previous work of rigid connected frames (1). The bilinear hysteresis loop shown in Fig. 3 is composed of two imaginary components, a linear component and an elasto-plastic component. The initial slopes of the hysteresis loop and of the linear and elasto-plastic components are k, k_1, and k_2, respectively, in which $k = k_1 + k_2$, $k_1 = pk$, $k_2 = qk$, and $p + q = 1$. The term p is the fraction of stiffness apportioned to the linear component, and q is the fraction apportioned to the elasto-plastic component. The second slope, k_1, of the hysteresis loop is the same as the initial slope of the linear component.

Based on Fig. 3, one can assume that an inelastic member has two equivalent components, and the incremental forces and deformations at the ends of these two components can be expressed in terms of stiffness coefficients as:

(a) linear component

$$\begin{Bmatrix} FM_p \\ FV_p \end{Bmatrix} = p[S_r] \begin{Bmatrix} \theta \\ Y \end{Bmatrix} \tag{11}$$

(b) elasto-plastic component

$$\begin{Bmatrix} FM_q \\ FV_q \end{Bmatrix} = q[S_r] \begin{Bmatrix} \Omega \\ Y \end{Bmatrix} \tag{12}$$

where $[S_r]$ is given in Eqs. (9b) through (9k) or Eqs. (10a) through (10i), $\{\theta\}=[\theta_i \; \theta_j]^T$ are the joint rotations, $\{\alpha\}=[\alpha_i \; \alpha_j]^T$, the plastic angles at the ends of the elasto-plastic component, and $\{\Omega\}=[\omega_i \; \omega_j]^T$, the end rotations of the elasto-plastic component. The moments and shears of the nonlinear member are a combination of the end forces of the components according to the state of yield. The force-deformation relationship associated with different state of yield conditions are discussed separately (4,5).

(a) <u>Both ends linear</u>: Since the member is still elastic, the plastic angles do not exist. Then $\alpha_i=\alpha_j=0$, $\omega_i=\theta_i$, and $\omega_j=\theta_j$. The force-deformation relationship of Eq. (9) or (10) is used.

(b) <u>i end nonlinear and j end linear</u>: Because the i end is nonlinear, the plastic angle α_i is not zero. Therefore, $\Delta\alpha_i=0$, $\Delta\alpha_j=0$, $\Delta\omega_i=\Delta\theta_i-\Delta\alpha_i$, and $\Delta\omega_j=\Delta\theta_j$. The incremental member forces are

$$\begin{Bmatrix} \Delta FM \\ \Delta FV \end{Bmatrix} = \begin{Bmatrix} \Delta FM_p \\ \Delta FV_p \end{Bmatrix} + \begin{Bmatrix} \Delta FM_q \\ \Delta FV_q \end{Bmatrix} \tag{13a}$$

or

$$\begin{Bmatrix} \Delta M_i \\ \Delta M_j \\ \Delta V_i \\ \Delta V_j \end{Bmatrix} = p[Sr] \begin{Bmatrix} \Delta\theta_i \\ \Delta\theta_j \\ \Delta y_i \\ \Delta y_j \end{Bmatrix} + q[Sr] \begin{Bmatrix} \Delta\theta_i-\Delta\alpha_i \\ \Delta\theta_j \\ \Delta y_i \\ \Delta y_j \end{Bmatrix} \tag{13b}$$

By using relations $\Delta M_i=\Delta M_{pi}$ and $p+q=1$, one can solve for $\Delta\alpha_i$ from Eq. (13b) as

$$\Delta\alpha_i = [1 \quad \frac{n}{m_1} \quad \frac{r_1}{m_1} \quad \frac{r_1}{m_1}] \; [\Delta\theta_i \; \Delta\theta_j \; \Delta y_i \; \Delta y_j]^T \tag{14}$$

Substituting Eq. (14) into Eq. (13b), the force-deformation relationship is derived as

$$
\begin{Bmatrix} \Delta M_i \\ \Delta M_j \\ \Delta V_i \\ \Delta V_j \end{Bmatrix}
=
\begin{bmatrix}
pm_1 & & & \text{Sym.} \\
pn & m_2 - q\dfrac{n^2}{m_1} & & \\
pr_1 & r_2 - q\dfrac{nr_1}{m_1} & s - q\dfrac{r_1}{m_1} & \\
pr_1 & r_2 - q\dfrac{nr_1}{m_1} & s - q\dfrac{r_1}{m_1} & s - q\dfrac{r_1}{m_1}
\end{bmatrix}
\begin{Bmatrix} \Delta\theta_i \\ \Delta\theta_j \\ \Delta y_i \\ \Delta y_j \end{Bmatrix}
\tag{15}
$$

    (c) <u>i end linear and j end nonlinear</u>: When the j end is nonlinear, $\Delta\alpha_i = 0$, $\Delta\alpha_j = 0$, $\Delta\omega_i = \Delta\theta_i$, and $\Delta\omega_j = \Delta\theta_j - \Delta\alpha_j$. Therefore,

$$
\begin{Bmatrix} \Delta M_i \\ \Delta M_j \\ \Delta V_i \\ \Delta V_j \end{Bmatrix}
= p[Sr]
\begin{Bmatrix} \Delta\theta_i \\ \Delta\theta_j \\ \Delta y_i \\ \Delta y_j \end{Bmatrix}
+ q[Sr]
\begin{Bmatrix} \Delta\theta_i \\ \Delta\theta_j - \Delta\alpha_j \\ \Delta y_i \\ \Delta y_j \end{Bmatrix}
\tag{16}
$$

Solve for $\Delta\alpha_j$ from the above equation by using $\Delta M_j = \Delta M_{pj}$ and $p+q=1$, one has

$$\Delta\alpha_j = [\frac{n}{m_2} \quad 1 \quad \frac{r_2}{m_2} \quad \frac{r_2}{m_2}] \; [\Delta\theta_i \; \Delta\theta_j \; \Delta y_i \; \Delta y_j]^T \tag{17}$$

Substituting Eq. (17) into Eq. (16), the following relationship can be derived

$$
\begin{Bmatrix} \Delta M_i \\ \Delta M_j \\ \Delta V_i \\ \Delta V_j \end{Bmatrix} = \begin{bmatrix} m_1 - q\dfrac{n^2}{m_2} & & \text{Sym.} & \\ pn & pm_2 & & \\ r_1 - q\dfrac{nr_2}{m_2} & pr_2 & s - q\dfrac{r_2}{m_2} & \\ r_1 - q\dfrac{nr_2}{m_2} & pr_2 & s - q\dfrac{r_2}{m_2} & s - q\dfrac{r_2}{m_2} \end{bmatrix} \begin{Bmatrix} \Delta\theta_i \\ \Delta\theta_j \\ \Delta y_i \\ \Delta y_j \end{Bmatrix}
\tag{18}
$$

(d) <u>Both ends nonlinear</u>: If both ends are nonlinear, the plastic angles, $\Delta\alpha_i$ and $\Delta\alpha_j$, are not zero. Then $\Delta\omega_i = \Delta\theta_i - \Delta\alpha_i$ and $\Delta\omega_j = \Delta\theta_j - \Delta\alpha_j$. The incremental forces are

$$
\begin{Bmatrix} \Delta M_i \\ \Delta M_j \\ \Delta V_i \\ \Delta V_j \end{Bmatrix} = p[Sr] \begin{Bmatrix} \Delta\theta_i \\ \Delta\theta_j \\ \Delta y_i \\ \Delta y_j \end{Bmatrix} + q[Sr] \begin{Bmatrix} \Delta\theta_i - \Delta\alpha_i \\ \Delta\theta_j - \Delta\alpha_j \\ \Delta y_i \\ \Delta y_j \end{Bmatrix}
\tag{19}
$$

By using the relations $\Delta M_i = \Delta M_{pi}$, $\Delta M_j = \Delta M_{pj}$ and $p+q=1$, one can solve for $\Delta\alpha_i$ and $\Delta\alpha_j$ as

$$
\begin{Bmatrix} \Delta\alpha_i \\ \Delta\alpha_j \end{Bmatrix} = \begin{bmatrix} 1 & 0 & \dfrac{r_1 m_2 - r_2 n}{m_1 m_2 - n^2} & \dfrac{r_1 m_2 - r_2 n}{m_1 m_2 - n^2} \\ 0 & 1 & \dfrac{r_2 m_1 - r_1 n}{m_1 m_2 - n^2} & \dfrac{r_2 m_1 - r_1 n}{m_1 m_2 - n^2} \end{bmatrix} \begin{Bmatrix} \Delta\theta_i \\ \Delta\theta_j \\ \Delta y_i \\ \Delta y_j \end{Bmatrix}
\tag{20}
$$

Substituting Eq. (20) into Eq. (19), the force-deformation relationship is derived as

$$
\begin{Bmatrix} \Delta M_i \\ \Delta M_j \\ \Delta V_i \\ \Delta V_j \end{Bmatrix} = \begin{bmatrix} pm_1 & & \text{Sym.} & \\ pn & pm_2 & & \\ pr_1 & pr_2 & ps - q\dfrac{N}{\ell} & \\ pr_1 & pr_2 & ps - q\dfrac{N}{\ell} & ps - q\dfrac{N}{\ell} \end{bmatrix} \begin{Bmatrix} \Delta\theta_i \\ \Delta\theta_j \\ \Delta y_i \\ \Delta y_j \end{Bmatrix}
\tag{21}
$$

When P-Δ effect is not considered, $N=0$.

3. Structural Formulation -- For a lumped mass system subjected to lateral forces, $\{F(t)\}$, the motion equation can be expressed as

$$
\begin{bmatrix} 0 & 0 \\ 0 & AM \end{bmatrix} \begin{Bmatrix} \ddot{x}_\theta \\ \ddot{x}_s \end{Bmatrix} + \begin{bmatrix} 0 & 0 \\ 0 & C \end{bmatrix} \begin{Bmatrix} \dot{x}_\theta \\ \dot{x}_s \end{Bmatrix} + \begin{bmatrix} K_{\theta\theta} & K_{\theta s} \\ K_{s\theta} & K_{ss} \end{bmatrix} \begin{Bmatrix} x_\theta \\ x_s \end{Bmatrix} = \begin{Bmatrix} 0 \\ F(t) \end{Bmatrix}
\tag{22}
$$

in which [AM] is the mass matrix; [C] is the damping matrix; $[\ddot{x}_\theta \ \ddot{x}_s]^T$, $[\dot{x}_\theta \ \dot{x}_s]^T$, and $[x_\theta \ x_s]^T$ are the acceleration, velocity, and displacement vectors, respectively; the subscripts θ and s represent structural joint rotation and sidesway, respectively; the geometric stiffness matrix is included in the system stiffness matrix when the P-Δ effect is considered. The time-dependent forcing function, $\{F(t)\}$, can be ground acceleration records for which $\{F(t)\}=-[AM]\{\ddot{x}_g\}$. Equation (22) can be condensed to the following form:

$$
[AM]\{\ddot{x}_s\} + [C]\{\dot{x}_s\} + [K_c]\{x_s\} = \{F(t)\}
\tag{23}
$$

where $[K_c]=[K_{ss}]-[K_{s\theta}][K_{\theta\theta}]^{-1}[K_{\theta s}]$.

The rotational displacements can be obtained by using the following formula

$$
\{x_\theta\} = -[K_{\theta\theta}]^{-1}[K_{\theta s}]\{x_s\}
\tag{24}
$$

4. Damping Property -- The following damping matrix is expressed in a linear combination of mass and stiffness such that the effect of inelastic behavior of both the joints and members on damping can be included.

$$
[C] = \alpha[AM] + \beta[K]
\tag{25}
$$

in which α and β are constants determined on the basis of the fundamental natural frequency, p_n, and the damping ratio, ρ, as follows: 1) mass plus stiffness proportional damping: $\alpha=\rho p_n$ and $\beta=\rho/p_n$, 2) mass proportional damping: $\alpha=2\rho p_n$ and $\beta=0$, and 3) stiffness proportional damping: $\alpha=0$ and $\beta=2\rho/p_n$.

5. Ductility Ratio -- Two definitions of the ductility ratio (3,5) were used to evaluate the ductility requirement of a bilinear system with a $2M_p$ elastic stress range. The first definition of ductility ratio is based on curvature which is defined in terms of Fig. 5 as follows:

$$
\mu_1 = \frac{\phi_{max}}{\phi_y} = 1 + \frac{\phi_o}{\phi_y}
\tag{26}
$$

in which ϕ_{max} is the maximum curvature ϕ_y curvature at yield, and ϕ_o

plastic curvature.

For the first half cycle, $M_{max}=M_p+\phi_o pk$ and $\phi_y=M_p/k$ from which Eq. (26) can be changed to the following form:

$$\mu_1 = 1 + \frac{M_{max}-M_p}{pM_p} \tag{27}$$

The plastic deformation for the second half cycle is $\phi_1=|M_2-M_1|/pk$, then

$$\mu_1 = 1 + \frac{|M_2-M_1|}{pM_p} \tag{28}$$

The second definition of the ductility ratio is the ratio of the dissipated strain energy of a member end to the total elastic strain energy in the member plus one as

$$\mu_2 = 1 + \frac{E_d}{E_e} \tag{29}$$

where E_d is the dissipated strain energy of a member end and E_e the total elastic strain energy in the member as shown in Fig. 6.

NUMERICAL PROCEDURES

Runge-Kutta Fourth-Order Method -- The Runge-Kutta fourth-order method has been widely used for the step by step integration (2) of the motion equation. To find the response at time $t+\Delta t$ by using this method, one is required to calculate the restoring force, $[K]\{x(t)\}$, of the system at time t. The restoring force must be calculated step by step as follows:

$$\{x(t+\Delta t)\} = \{x(t)\} + \Delta t\{\dot{x}(t)\} + \frac{\Delta t}{6}(\{a\} + \{b\} + \{d\}) \tag{30}$$

$$\{\dot{x}(t+\Delta t)\} = \{\dot{x}(t)\} + \frac{1}{6}(\{a\} + 2\{b\} + 2\{d\} + \{e\}) \tag{31}$$

$$\{\ddot{x}(t+\Delta t)\} = \{\ddot{x}(t)\} + [AM]^{-1}(\{F(t+\Delta t)\} - \{F(t)\}$$
$$- [K-K_g](\{x(t+\Delta t)\}-\{x(t)\}) - [C](\{\dot{x}(t+\Delta t)\}-\{\dot{x}(t)\})) \tag{32}$$

where
$\{\Delta x(t)\} = \{x(t)\} - \{x(t-\Delta t)\}$,
$\{\Delta\dot{x}(t)\} = \{\dot{x}(t)\} - \{\dot{x}(t-\Delta t)\}$,
$\{a\} = \Delta t[AM]^{-1}(\{F(t)\} - \{R(t)\})$,
$\{b\} = \Delta t[AM]^{-1}(\{F(t+\frac{\Delta t}{2})\} - \{R(t)\} - \frac{\Delta t}{2}\{R1(t)\} - \frac{1}{2}[C]\{a\})$
$\{d\} = \Delta t[AM]^{-1}(\{F(t+\frac{\Delta t}{2})\} - \{R(t)\} - \frac{\Delta t}{2}\{R1(t)\} - \frac{\Delta t}{4}[K-K_g]\{a\}$
$\quad -\frac{1}{2}[C]\{b\})$,
$\{e\} = \Delta t[AM]^{-1}(\{F(t+\Delta t)\} - \{R(t)\} - \Delta t\{R1(t)\} - \frac{\Delta t}{2}[K-K_g]\{b\}$
$\quad - [C]\{d\})$,
$\{R(t)\} = \{R(t-\Delta t)\} + [K-K_g]\{\Delta x(t)\} + [C]\{\Delta\dot{x}(t)\})$, and
$\{R1(t)\} = \{R1(t-\Delta t)\} + [K-K_g]\{\Delta\dot{x}(t)\}$.

NUMERICAL EXAMPLES

    The one story unbraced steel frame shown in Fig. 7 was used as a
model to study the effects of P-Δ and semi-rigid connections on the
response behavior. The height of the steel frame is 10 feet (3.05 m),
and the girder length is 20 feet (6.10 m). The lumped mass of the floor
level is 0.0002 k-sec^2/in. (35.03 kg). The material property, EI, for
all members is 1000 ksi (689.48 kN/cm^2), the plastic moment, M_p, is
0.2 k-in. (22.60 N-m), and the coefficients p and q are 0.05 and 0.95
respectively. The frame is subjected to two loading conditions: one is
a cyclic load of 0.02 sin(πt) kips (88.96 sin(πt) N) acting
on the first floor level and the other is due to the base excitation of
the N-S component of the 1940 El Centro earthquake. The results are
discussed separately.
    1. Effects of Semi-rigid Modeling on Response Behavior -- Two
different models of semi-rigid connections were used for the purpose of
comparison. The first model is assumed to be linear for the M-θ' curve,
and the second is assumed to be bilinear. The percentage of rigidity,
τ, is 0.8 for the linear range, and is 0.333 for the plastic range. In
the bilinear model, the transition moment, M_t, is 0.25 k-in. (28.245
N-m) for the cyclic loading case, and 0.23 k-in. (25.985 N-m) for the
seismic analysis. The amplification factor for the 1940 El Centro
earthquake records is 1.3. The P-Δ effect is considered with an axial
force N=0.015 kips (44.482 N) was applied on the top of each column.
Damping is not considered in the analysis.
    The results are plotted in Figs. 8 through 13. As shown in Figs.
8 and 11, the maximum rotation is significantly increased when the bilinear model of semi-rigid connection is considered. The main reason
for this increase is that the stiffnesses of the connections are reduced
when the transition moment is reached; consequently the system stiffness
is reduced. Figures 9 and 12 show the relationships between moments and
slip-angles at the beam-column connections in which the bilinear model
yields larger deformations. The lateral displacements corresponding to
these two loading cases are shown in Figs. 10 and 13 where the bilinear
model of semi-rigid connections produced larger displacements.
    The maximum ductility ratios for the seismic analyses are shown
in Table I where the ductilities of the columns are developed at the
supports and the ductilities of the girder are developed at the member
ends. The ductilities of the members are higher than those of the connection because M_t is greater than M_p.

    2. P-Δ Effects -- The beam-column connections are behaved as
linear with the rigidity coefficient τ=0.8. The amplification factor
for the earthquake record is 1.0. The axial force applied on the top
of each column is changed as 0, 0.01 kips (44.482 N), and 0.021 kips
(93.412 N), respectively.
    Figures 14 and 15 show the displacement response histories. These
figures show that the responses have been increased when the axial
forces were increased. The incremental responses in the plastic range
are greater than that in elastic range. It is worth noting that the
maximum allowable axial force, N, is 0.021 kips (93.412 N) which is
5.357% of the elastic buckling load 0.392 kips (1.744 kN) of the semi-
rigid connected system. When N is greater than 0.021 kips, the structure

becomes unstable. It is apparent that if a bilinear model of the semi-rigid joint is used, larger displacements will be obtained.

As shown in Table II, the ductility ratio at the column is in-creased when the axial force is increased, however, the ductility ratio in the girder is decreased.

3. Damping Effects -- The damping ratio is changed as 0, 5%, and 10% for which the damping matrix is evaluated on the basis of mass proportional damping. The axial force N, is 0.01 kips (44.482 N).

Figures 16 and 17 show the maximum displacement is reduced when the damping ratio is increased. The ductility ratios are shown in Table III and are reduced for both the columns and the girder when damping is considered.

CONCLUSIONS

General stiffness coefficients of a typical elastic and inelastic beam-column member are developed with consideration of linear and bi-linear semi-rigid connection with and without P-Δ effects. Numerical procedures of incremental analysis based on the Runge-Kutta fourth order method are also presented.

A structural model is used to study the effect of the P-Δ force and that of semi-rigid connections on the response behavior of a structure subjected to cyclic loading and seismic excitations. The response behavior is expressed in terms of the lateral displacements, joint rotations, slip angles, and ductility ratios. The examples show that the displacement response is significantly increased when the beam-column connections behave bilinearly, and the ductility ratios are increased. The P-Δ effect significantly increases the structural response, and the buckling load of the semi-rigid model is much less than that of the rigid-connected model. The ductility ratios based on the energy formu-lation are slightly higher than those based on the curvature formulation, but a similar trend for both approaches can be observed.

REFERENCES

1. Cheng, F.Y., and Oster B.K., "Ultimate Instability of Earthquake Structures", Journal of the Structural Division, ASCE, Vol. 102, pp. 961-972, 1976.
2. Cheng, F.Y., and Botkin, M.E., "Second-Order Elasto-Plastic Analysis of Tall Buildings with Damped Dynamic Excitations", Pro. on the Finite Element Method in Civil Engineering, McGill University, Montreal, Canada, pp. 549-564, 1972.
3. Cheng, F.Y., Computer Methods in Structural Dynamics and Earthquake UMR, 1984.
4. Cheng, F.Y., and P. Kitipitayangkul, Investigation of the Effect of 3-D Parametric Earthquake Motions on Stability of Elastic and Inelastic Building Systems, Final Report prepared for the National Science Foundation. Available at the U.S. Department of Commerce, National Technical Information Service, Springfield, VA 22151, PB80-176936, (392 pages).
5. Cheng, F.Y., and Oster, B.K., Dynamic Instability and Ultimate Capacity of Inelastic Systems Parametrically Excited by Earthquake -- Part II, Technical Report of the National Science Foundation. National Technical Information Service, U.S. Department of Commerce, PB261097/AS, 1976.

6. Frye, M.J. and, Morris, G.A., "Analysis of Flexibly Connected Steel Frames", _Canadian Journal of Civil Engineering_, Vol. 2, 1975.

7. Jones, W.S., Kirby, P.A., and Nethercot, D.A., "Columns with Semi-rigid Joints", _Journal of the Structural Division_, ASCE, Vol. 108, No. ST2, pp. 361-372, 1982.

8. Helmut Krawinkler, and Egor P. Popov, "Seismic Behavior of Moment Connections and Joints", _Journal of the Structural Division_, ASCE, Vol. 108, No. ST2, Feb., 1982, pp. 373-391.

9. Kirit V. Patel, and W. F. Chen, "Nonlinear Analysis of Steel Moment Connections", _Journal of the Structural Division_, ASCE, Vol. 110, No. ST8, Aug., 1984, pp. 1861-1874.

10. Piotr D. Moncarz, and Kurt H. Gerstle, "Steel Frames with Non-linear Connections", _Journal of the Structural Division_, ASCE, Vol. 107, No. ST8, Aug., 1981, pp. 1427-1441.

11. Romstad, K. M., and Subramanian, C. V., "Analysis of Frames with Partial Rigidity", _Journal of the Structural Division_, ASCE, Vol. 96, No. ST11, Nov., 1970, pp. 2283-2300.

12. Tuma, J.J., and Cheng, F.Y., _Dynamic Structural Analysis_, McGraw-Hill Book Company, 1983.

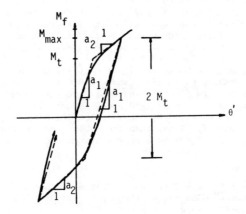

Fig. 1 Moment v.s. Slip-angle of a Semi-Rigid Joint.

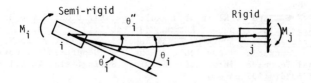

Fig. 2 Deformation between Member-end and Connection.

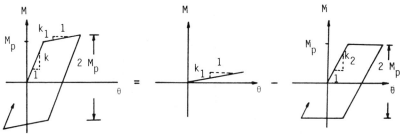

(a) Bilinear Hysteresis   (b) Linear Component   (c) Elasto-Plastic
    Loop   Component

Fig. 3   Bilinear Model for Members.

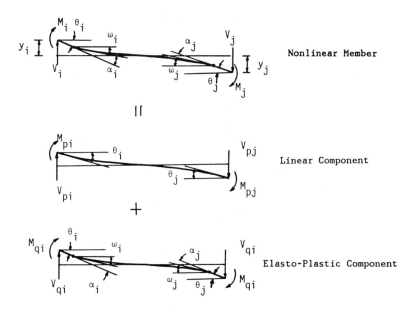

Fig. 4   Bilinear Member Components.

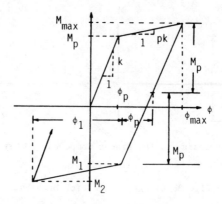

Fig. 5   Ductility Based on Moment Curvature.

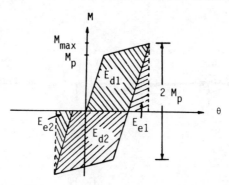

Fig. 6   Ductility Based on Energy.

Semi-rigid Connections

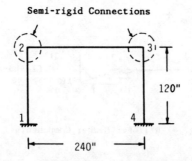

Fig. 7   One-Story Structural Model.
          (1 in.=2.54 cm, 1 kip=4.448 kN).

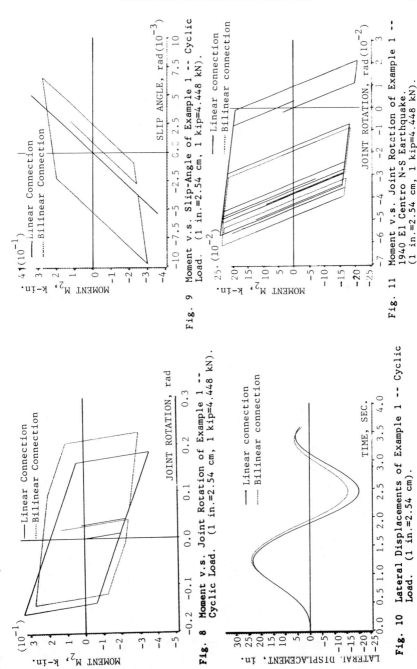

Fig. 8 Moment v.s. Joint Rotation of Example 1 -- Cyclic Load. (1 in.=2.54 cm, 1 kip=4.448 kN).

Fig. 9 Moment v.s. Slip-Angle of Example 1 -- Cyclic Load. (1 in.=2.54 cm, 1 kip=4.448 kN).

Fig. 10 Lateral Displacements of Example 1 -- Cyclic Load. (1 in.=2.54 cm).

Fig. 11 Moment v.s. Joint Rotation of Example 1 -- 1940 El Centro N-S Earthquake. (1 in.=2.54 cm, 1 kip=4.448 kN).

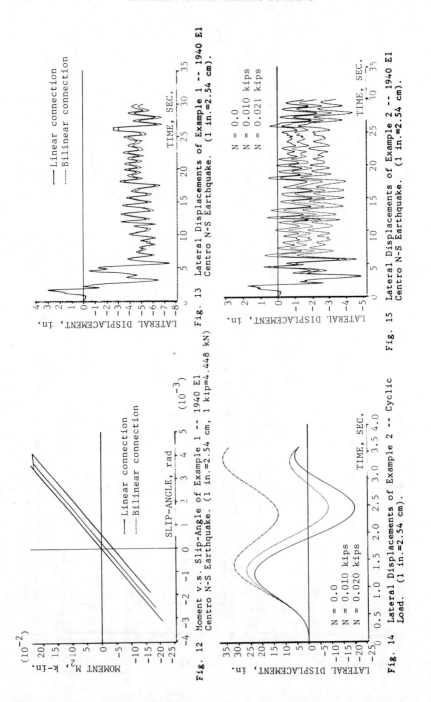

Fig. 12 Moment v.s. Slip-Angle of Example 1 -- 1940 El Centro N-S Earthquake. (1 in.=2.54 cm, 1 kip=4.448 kN).

Fig. 13 Lateral Displacements of Example 1 -- 1940 El Centro N-S Earthquake. (1 in.=2.54 cm).

Fig. 14 Lateral Displacements of Example 2 -- Cyclic Load. (1 in.=2.54 cm).

Fig. 15 Lateral Displacements of Example 2 -- 1940 El Centro N-S Earthquake. (1 in.=2.54 cm).

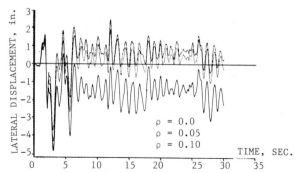

Fig. 16   Lateral Displacements of Example 3 -- Cyclic
Load.   (1 in.=2.54 cm).

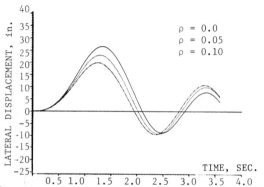

Fig. 17   Lateral Displacements of Example 3 -- 1940 El
Centro N-S Earthquake.   (1 in.=2.54 cm).

Table I.   Maximum Ductility Ratios of the Seismic Analysis in

Example 1.

Connection	Based on Curvature Ductility ratio, $(\mu_1)_{max}$			Based on Energy Ductility Ratio, $(\mu_2)_{max}$		
Model	Column	Girder	Connection	Column	Girder	Connection
Bilinear	12.727	5.146	1.189	11.989	6.221	1.341
Linear	12.864	5.421	1.0	12.100	6.251	1.0

Table II. Maximum Ductility Ratios of the Seismic Analysis in
           Example 2. (1 kip = 4.448 kN)

Axial Force	Based on Curvature Ductility ratio, $(\mu_1)_{max}$		Based on Energy Ductility Ratio, $(\mu_2)_{max}$	
(kips)	Column	Girder	Column	Girder
N = 0.	8.306	3.216	9.014	3.560
N = 0.01	8.418	3.201	9.030	3.524
N = 0.021	8.590	3.193	9.195	3.498

Table III. Maximum Ductility Ratios of the Seismic Analysis in
            Example 3.

Damping Ratio ρ	Based on Curvature Ductility ratio, $(\mu_1)_{max}$		Based on Energy Ductility Ratio, $(\mu_2)_{max}$	
	Column	Girder	Column	Girder
0	8.418	3.201	9.030	3.524
0.05	7.307	2.837	8.798	3.078
0.10	7.198	2.430	8.558	2.975

BIDIRECTIONAL SEISMIC RESPONSE OF SIMPLE STRUCTURES

C.R. Thewalt[1], S.A. Mahin[2]

For design purposes it has been common to consider that seismic excitations act in one direction at a time. Analytical studies over the past decade have indicated that this may not be an adequate assumption in certain circumstances. In particular, coupled torsional lateral response and response due to multiple components of base excitation can in some cases lead to significantly increased response. Many of these analyses have assumed linear elastic structural behavior. Studies that have used inelastic models indicate that such structures may behave much differently than structures that respond elastically. To help clarify this difference a series of experimental and analytical studies have been initiated focusing on both torsional response and response to multicomponent base excitation.

INTRODUCTION

In evaluating the design implications of the available analytical studies on inelastic torsional and biaxial structural response (Bozorgnia 1984; Fujiwara 1982, 1984; Gergeley 1984; Padilla-Mora 1974; Tso 1984), it is unclear how much of the predicted behavior is due to the simplifications introduced in the analytical models. Experimental results are needed to see if predicted behavior actually occurs and also to provide a basis for the verification of analytical models. To this end an extensive experimental and analytical investigation has been planned. The objectives of these long range studies are to: (1) obtain basic experimental data in the inelastic range on coupled torsional-translational response and on response to multiple components of base excitation; (2) to validate and, if necessary, improve analytical models for predicting this behavior, and (3) to perform analytical studies to develop design recommendations for the effective seismic-resistant design of structures taking these factors into account. This paper describes the experimental procedure and preliminary results obtained to date. Future work is also discussed.

To obtain the required data it would be most realistic to design a specimen and test it dynamically on a shaking table. There are, however, no tables available in the U.S. that can subject a reasonably sized specimen to multiple lateral ground motion inputs. The pseudo-dynamic test method (PDTM) (Shing 1984) could in principle be used to apply any general fixed base excitation to a specimen, but the method has previously been used only on planar structures subjected to single component base excitation (Mahin 1985). Thus, in these studies the PDTM has been extended to consider three dimensional structural response under multiple components of excitation.

To verify the ability of the PDTM to simulate three dimensional response under multiple ground motion inputs, as well as to obtain basic data on inelastic dynamic response, a series of tests was performed on a shaking table for a torsionally excited structure subjected to a single component of excitation. The specimen was designed with a large eccentricity with respect to both its major axes, so that even the single component of base excitation considered would induce considerable torsional response and biaxial column bending. Pseudo-dynamic tests were then performed using the measured shaking table acceleration histories as input. These tests show that the PDTM provides very similar response in both the elastic and inelastic ranges. Thus, subsequent tests will use the PDTM to investigate behavior of

1. Research Assistant, University of California, Berkeley, CA.
2. Professor of Civil Engineering., University of California, Berkeley, CA.

symmetrical and unsymmetrical specimens under single as well as multiple components of ground motion input.

SHAKING TABLE TESTS

The objective in designing the initial shaking table specimen was to make it eccentric with respect to both principal axes, so that the single ground motion component from the table would result in as much lateral-torsional interaction as possible. In order to simplify analytical modeling and correlation, a single story, one bay by one bay, steel frame was selected. The specimen is shown in Fig 1. The upper deck was rigid relative to the four corner columns. One of the four corner columns was rotated 90 degrees about a vertical axis relative to the others to provide a stiffness eccentricity with respect to both the major axes of the upper deck. The columns are 1.21 m (48 inches) in length and are standard S3x7.5 sections in A36 steel. They were selected to give all natural periods in the 0.15 sec. to 0.35 sec. range for the deck masses of 44.5 kN (10 kips) and 62.3 kN (14 kips). Two different masses were employed in the tests as a simple means of testing two slightly different structures. The specimen was mounted on the shaking table at an angle of 45 degrees relative to the direction of motion. This helped to induce torsional response, and effectively provide two correlated components of ground motion input (each equal to 0.7071 of the measured table acceleration).

The ground motions considered were the NS component of the 1940 El Centro record with peak acceleration varied between 0.13 g and 1.6 g and the S74W component of the 1971 Pacoima Dam record with a peak acceleration of 0.8 g. A total of eight tests were run with two resulting in elastic response and the remainder resulting in increasing levels of inelastic behavior. Representative results for an elastic and an inelastic test are shown in Figs. 2, 3. Specimens exhibited substantial biaxial response as expected. In addition, torsional response was also substantial, as can be seen in the results.

PSEUDO-DYNAMIC TESTS

The form of the pseudo-dynamic test method used to simulate multiple components of base excitation is very similar to the algorithm used for planar structures with single component excitation (Mahin 1985, Shing 1984). Only the right hand side of the equations of motion need be changed to account for the loading from each ground motion component.

The time discretized equations of motion for time step i, as used in pseudo-dynamic testing, are given by:

$$\mathbf{M} \, \mathbf{a}_i + \mathbf{C} \, \mathbf{v}_i + \mathbf{r}_i - \mathbf{K}_G \, \mathbf{d}_i = \mathbf{M} \, \mathbf{B} \, \mathbf{a}_{g_i} \qquad (1)$$

where
$\mathbf{M}, \mathbf{C}$	:	mass and damping matrices;
$\mathbf{K}_G$	:	geometric stiffness matrix;
$\mathbf{a}_i, \mathbf{v}_i, \mathbf{d}_i$	:	acceleration, velocity and displacement vectors;
$\mathbf{r}_i$	:	measured structural restoring force vector;
$\mathbf{B}$	:	ground acceleration transformation matrix;
$\mathbf{a}_{g_i}$	:	ground acceleration vector.

The component $\mathbf{B}_{ij}$ is the acceleration at structural degree of freedom i when the structure acts as a rigid body under a unit acceleration for ground component j. In a planar test with a single horizontal ground motion component, $\mathbf{B}$ becomes the familiar unit vector. In a test, $\mathbf{M}, \mathbf{C}, \mathbf{K}_G, \mathbf{B}$ and the ground acceleration history are defined by the user. At each step, $\mathbf{d}_{i+1}$ is computed based on the user supplied data and the measured restoring force vector, using an appropriate numerical integration strategy (Shing 1984). The restoring force is then measured for use in the next step's calculations.

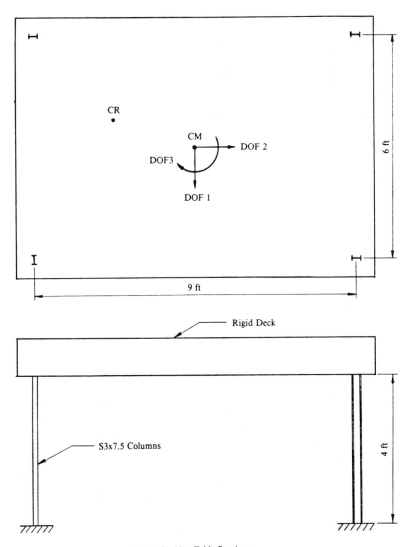

Fig. 1   Shaking Table Specimen

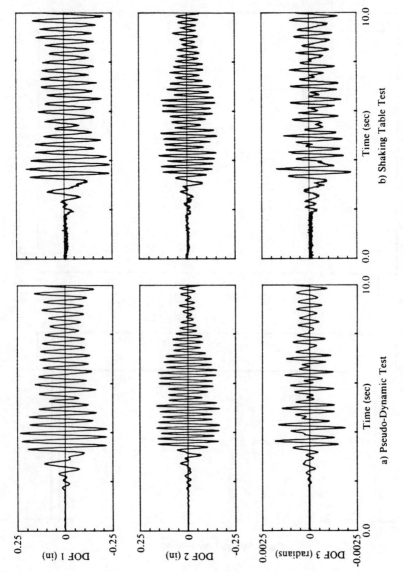

Fig. 2 Shaking Table and Pseudo-Dynamic Elastic Level Displacement Response

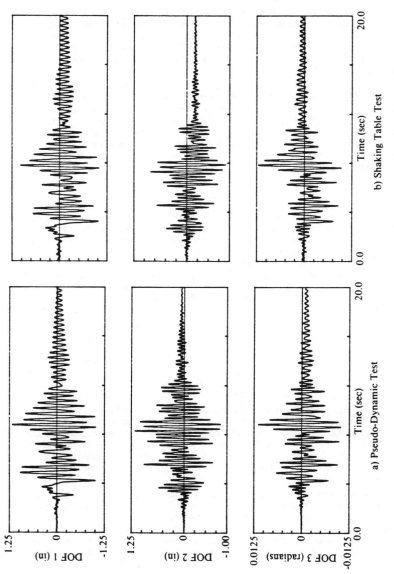

Fig. 3 Shaking Table and Pseudo-Dynamic Inelastic Level Displacement Response

The user supplied geometric stiffness matrix is only used in tests where it is not necessary to include the full design weight on the specimen. This may be appropriate in certain pseudo-dynamic tests. However, previous studies on structural response under multiple components of base excitation as well as torsional/translational response of structures have shown the P-delta effect has a very important influence. Thus, the actual mass is included in all of these pseudo-dynamic tests.

Trial pseudo-dynamic tests of a multiple degree of freedom structure demonstrated the necessity of removing experimental errors. Early tests were plagued by cumulative errors from poor transducers and poor hydraulic control. By employing better high performance test equipment and instrumentation, it was possible to reduce the errors and the pseudo-dynamic response histories were very similar to those measured on the shaking table, as shown in Figs. 2, 3. Even the elastic pseudo-dynamic tests, where the structural response is most likely to be contaminated by cumulative experimental errors, appeared to have very good correlation with the shaking table results.

CURRENT TEST PROGRAM

In the subsequent experimental studies the response of symmetric and unsymmetric structures subjected to uniaxial and biaxial ground motions are to be compared. However, the model from the verification study cannot be used. Rotating one column to make the structure symmetric/unsymmetric will make direct comparison of the results for symmetrical and unsymmetrical structures difficult, since the diagonal terms as well as the off-diagonal terms of the stiffness matrix change. These stiffness matrix changes will result in shifted natural frequencies and also in substantially different response, especially for the lightly damped elastic case. In addition, the eccentricity and frequency ratios of the specimen are outside of the bounds of values considered in previous analytical studies (Bozorgnia 1986, Tso 1984).

The eccentricity of the shaking table model is very large also, well beyond the reasonable range for structures. It is essentially equal to the mass radius of gyration. Also, it would be desirable to have the eccentricity occur along one principal axis only, in order to make comparison with available analytical studies possible. A better specimen would combine both a smaller eccentricity than the initial shaking table model, and would guarantee that all natural frequencies remain equal for the symmetrical and unsymmetrical models.

The design of such a model is in fact simple, as shown in Fig 4. By moving the two column lines, any desired eccentricity can be achieved. In this specimen, the longitudinal direction remains uncoupled in the elastic range with respect to both the lateral and torsional degrees of freedom. The longitudinal frequency will not change, and for any desired eccentricity the column offsets can be adjusted to ensure that the diagonal elements of the stiffness matrix remain the same, or that one of the two remaining natural frequencies does not change.

Tests are currently underway, using scaled versions of El Centro and Taft earthquake records. Elastic and inelastic response of the two models are being studied using both one component and two component earthquakes. The tests will give experimental data that covers the planar case (symmetric structure with uniaxial ground motion), the biaxial case (symmetric structure with biaxial ground motion), the torsionally coupled case (eccentric structure with uniaxial ground motion), and the combined case of torsional response under biaxial ground motion input. These tests are in progress and the results will be presented subsequently.

ACKNOWLEDGEMENTS

The work performed herein was funded by the National Science Foundation and also by a graduate fellowship from MTS Systems, Inc. This support is gratefully acknowledged.

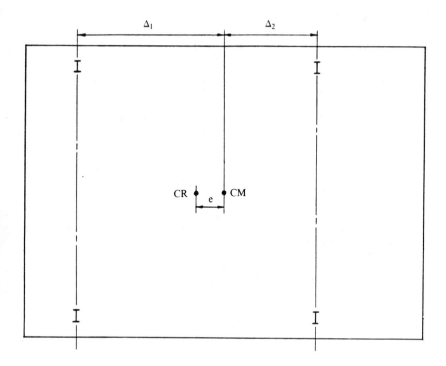

Fig. 4   Torsional/Biaxial Specimen

Results, findings and opinions, however, are those of the authors.

REFERENCES

Bozorgnia, Y., Tso, W.K., "Inelastic Earthquake Response of Monosymmetric Structures", *Journal of Structural Engineering,* ASCE, Feb., 1986.

Fujiwara, T., Kitahara, A., "Dynamic Failure Tests of Space Structures Subjected to Bi-Directional Horizontal Ground Motion", *Proceedings of the Sixth Japan Earthquake Engineering Symposium,* 1982.

Fujiwara, T., Kitahara, A., "On the Aseismic Safety of Space Structures Under Bi-Directional Ground Motion", *Eighth World Conference on Earthquake Engineering,* San Francisco, CA, 1984, Vol. IV, pp. 639-646.

Gergeley, P., Bejar, L.A., "Non-Linear Response of Torsionally Coupled Buildings for Multicomponent Earthquake Excitations", *Eighth World Conference on Earthquake Engineering,* San Francisco, CA, 1984, Vol. IV, pp. 227-234.

Mahin, S.A., and Shing, P.B., "Pseudo-Dynamic Method for Seismic Testing," *Journal of Structural Engineering,* ASCE, July 1985.

Padilla-Mora, R., Schnobrich, W.C., "Non-Linear Response of Framed Structures to Two-Dimensional Earthquake Motion", *Structural Research Series No. 408,* University of Illinois at Urbana-Champaign, Urbana, Illinois, 1974.

Shing, P.B., and Mahin, S.A., "Pseudo-Dynamic Test Method for Seismic Performance Evaluation: Theory and Implementation," *UCB/EERC-84/01,* Earthquake Engineering Research Center, University of California, Jan. 1984.

Tso, W.K., Sadek, A.W., "Inelastic Response of Eccentric Buildings Subjected to Bi-Directional Ground Motions", *Eighth World Conference on Earthquake Engineering,* San Francisco, CA, 1984, Vol. IV, pp. 203-210.

STUDY OF THE COLLAPSE OF A REINFORCED CONCRETE
MODEL STRUCTURE UNDER SIMULATED EARTHQUAKE MOTION

Arturo E. Schultz[*], A.M. ASCE

The collapse, under simulated earthquake motion, of a
multistory r/c frame structure with columns flexurally
weaker than beams (SBWC) is studied. A small-scale model,
comprising parallel nine-story, three-bay frames, was
constructed and tested on an earthquake simulator.
Structural response is summarized over a series of simulated
earthquakes of increasing intensity. Displacement records,
during the simulated earthquake that produced collapse, are
incomplete. Different methods for correcting and
integrating acceleration records to produce displacement
response are considered.

INTRODUCTION

The collapse of an instrumented multistory structure subjected to
strong base motions in a controlled laboratory environment is a rare
occurrence. Such an event provides an invaluable opportunity to study
response upon impending collapse. This paper presents test results of
the behavior upon collapse of a small-scale, nine-story reinforced
concrete (r/c) frame structure that was tested on an earthquake
simulator.

This test is part of a broader investigation of the earthquake
response of r/c frame structures with columns that are flexurally
weaker than the beams (Schultz, 1986). This type of structure is
susceptible to collapse at large displacements and column strength
deterioration under load reversals. For this reason, the engineering
community discourages the use of "strong beam"-"weak column" (SBWC)
frame structures in areas of high seismic risk. However, the SBWC
frame configuration is permitted in many areas of moderate seismic risk,
and because of their architectural functionality they are frequently
used. This project was undertaken to fill the void of experimental
data on the earthquake response of frames of this type.

The tests were carried out at the University of Illinois at Urbana-
Champaign by the author, under supervision of Professor Mete A. Sozen.
Financial support was provided by the National Science Foundation under
research grant CEE 81-14977.

DESCRIPTION OF EXPERIMENT

The test specimen was modelled after the SBWC frame concept. The
model structure was planar in configuration and comprised two parallel

[*]Assistant Professor, Department of Civil and Mechanical Engineering,
Southern Methodist University, Dallas, Texas, 75275.

frames which were coupled by rigid diaphragms at each story (Fig. 1).
Each frame had three bays and nine stories, of which the first story
was 40% taller than the others. Dimensions were selected as typical
for SBWC construction assuming a length scale factor of 1:15 (Fig. 1).
Added mass was provided by the story diaphragms at a rate of 1.14 kips
(5.07 kN) per story for a total structure weight of 10.25 kips
(45.6 kN). Fixed base conditions were simulated by rigid base girders.

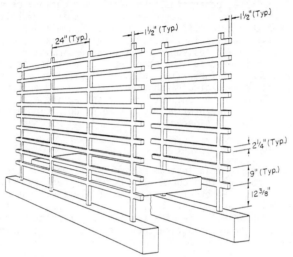

Fig. 1   Test Structure

The specimen was designed to resist a simulated version of the NS
component of El Centro 1940 along its major horizontal direction.   The
time scale was compressed by a factor of 2.5 to shift the frequency
content of the ground motion to the frequency range of the test specimen.
Base accelerations were scaled to a peak value of 0.35 g. Design forces
were obtained assuming a reduced-stiffness linear model of the structure
(Shibata and Sozen, 1976). Gross-section member stiffnesses were
reduced to simulate nonlinear response, and an increased equivalent
damping factor was used to account for hysteretic energy dissipation
with nonlinear deformation. The reduced-stiffness linear model was used
in a modal response-spectrum analysis to determine minimum design forces.

The test structure was constructed of microconcrete and model
reinforcement. The microconcrete was mixed in the laboratory from sand,
gravel, and Type III Portland cement, and had a compression strength of
5400 psi (37 N/mm^2). Longitudinal reinforcement for the beams was No. 7
gage wire and No. 13 gage wire was used to reinforce the columns. Both
sizes of wire had a yield stress of 56 ksi (386 N/mm^2), and strengths
were 63 and 61 ksi (435 and 420 N/mm^2) for the beam and column steel,
respectively.

The test structure was reinforced to satisfy the minimum
requirements of the design analysis. The first-story columns had a

total reinforcement ratio of 2.9%. The lateral strength of the columns was reduced over structure height to meet the design requirements. Beams were provided enough longitudinal reinforcement so that the sum of the flexural strengths of the beams at a joint exceeded the sum of the column flexural strengths. Sufficient transverse reinforcement was provided to all members to prevent yielding of the transverse hoops. To minimize joint distress, the joints were reinforced with steel spirals, and longitudinal reinforcement was run continuous through the joints.

The structures were tested on the University of Illinois Earthquake Simulator (Sozen et. al., 1969). Response measurements during the simulations included displacements and accelerations at each level, which were measured with displacement transducers (LVDT) and accelerometers, respectively. The structure was subjected to a series of four simulated earthquakes. The first motion was the "design" earthquake, which was followed by a repetition of this motion. Peak base accelerations for the last two runs were nominally 1.5 and 4 times as large as that of the "design" earthquake (0.35 g). Low-amplitude free-vibration tests to determine dynamic properties preceded all simulations. The model structure collapsed after a few cycles during the fourth or "collapse" run. This paper focuses on the response to the last simulated earthquake, but a brief summary of response during the first three runs is provided.

SUMMARY OF RESPONSE

The peak base acceleration during the initial simulation was equal to the target value of 0.35 g (Fig. 2). The specimen responded to this motion with a maximum top acceleration of 0.9 g, corresponding to an amplification factor of 2.5. The base shear force history was obtained as the sum over structure height of story inertia forces, which were computed as the product of lumped story mass and measured acceleration. A maximum value of base shear of 3.9 kips (17.3 kN) was determined for the first run (Fig. 2). The corresponding base shear coefficient, assuming a total structure weight of 10.25 kips (45.6 kN), was 0.38.

The top and first-story displacement histories are shown in Fig. 2. The structure resisted the inertia loads with a maximum top displacement of 1.0 in. (25 mm), or 1.2% of total frame height. The first-story displacement history is quite similar to that of the top level, indicating a dominant first mode. Maximum first-story displacement was 0.38 in. (9.7 mm), which represents 40% of the maximum top displacement and 3.1% of the total first-story height.

Yielding of the first-story columns occurred during the "design" earthquake and significantly affected response. First-story force-displacement response during the first 4 sec. of the "design" motion is shown in Fig. 3. The first-story columns yielded in both loading directions at a deflection of 0.17 in. (4.3 mm). Yielding of the story is defined as a marked deviation from a linear force-displacement relation. The first story was loaded to a maximum nonlinear deformation of 0.38 in. (9.7 mm) which produced a permanent deformation of 0.07 in. (2 mm).

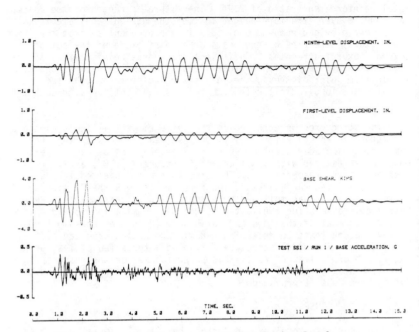

Fig. 2   Measured Response Histories During Run 1

The effect on response of the damage sustained during the first run is best illustrated by the change in apparent first-mode frequency. A low-amplitude free-vibration test before the simulation indicated an apparent first-mode frequency of 3.9 Hz. The frequency associated with the apparent period during the cycle of maximum response was 2.3 Hz, and at the end of the simulated earthquake the dominant period of the waveform was 1.9 Hz.

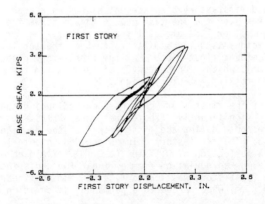

Fig. 3   First-Story Force-Displacement Response During Run 1

The second simulated earthquake was intended to be a repetition of the design motion, however, the actual motion was somewhat less intense than that of the first run. The Housner spectrum intensity (Housner, 1959), modified to reflect the time compression of the base motion, was 5% smaller for the second run than for the first simulated earthquake. Structural response during the second run did not exceed the maximum values observed during the "design" run.

The third run was nominally 1.5 times as intense as the first run. The maximum acceleration recorded at the base of the structure was 0.53 g. Maximum top acceleration was 1.36 g, for an acceleration amplification factor of 2.6. Top and first-level displacement maxima were 1.25 and 0.53 in. (31.8 and 13.5mm), respectively. These drifts represent 1.5 and 4.3% of total and first-story heights, respectively. The first story underwent loading into the nonlinear range during three half-cycles of response, and this further softened the structure. Apparent first-mode vibration frequencies were 1.8 and 1.6 Hz during the maximum response cycle and at the end of the base motion, respectively.

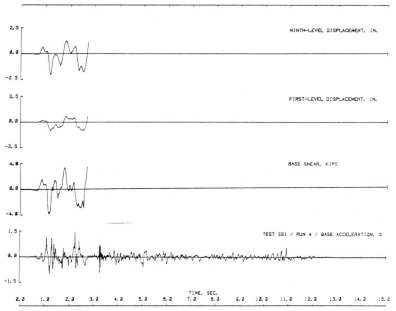

Fig. 4   Measured Response Histories During Run 4

The earthquake simulator was operated at maximum capacity during the last simulation, for the purpose of observing structural response in the vicinity of collapse. The resulting base motion had a maximum base acceleration of 1.44 g, or roughly four times that of the "design" motion. The structure responded to the first 2.75 sec. of the simulated earthquake motion in a stable fashion, for approximately four cycles. Response histories during this interval are shown in Fig. 4, along with the measured base acceleration record for the duration of the event.

The LVDT signals saturated the data acquisition computer system at a time of 2.75 sec., because the displacement response of the model structure was very large. At a time of 3.0 sec. the cores were pulled out of the displacement transducers. The high-frequency "spike" in the base acceleration record indicates that collapse occurred at a time of 3.3 sec. (Fig. 4). On the basis of observed response during the first three simulated earthquakes, the maximum-response cycle would be expected during the third 1-sec. interval of time. From 8-mm and 16-mm films of the "collapse" run, the first story of the structure was observed to deform to very large displacements until it became unstable under the column axial loads and collapsed. Large column axial loads, induced by impact against the simulator platform, resulted in the subsequent collapse of the sixth story.

Fig. 5   First-Story Force-Displacement Response During Run 4

Maximum top acceleration during the stable portion of response to the last simulated earthquake was 1.75 g, corresponding to a rather anemic acceleration amplification factor of 1.2. The first story had softened considerably and appeared to move in somewhat close unison with the ground. Maximum displacements at the top and first levels during first 2.75 sec. of the run were 2.06 and 0.86 in. (52.3 and 21.8 mm), respectively. Inspection of the first-story force-displacement relation (Fig. 5) reveals a large excursion into the nonlinear range during the first cycle, at a time of 1.2 sec. Base shear at this time was 3.8 kips (16.9 kN), which generated a displacement of 0.86 in. (21.8 mm) in the first story for a first-story drift ratio of 7%. During the next two half-cycles the first story developed consecutively smaller base shear forces of 3.4 and 2.9 kips (15.1 and 12.9 kN), respectively.

RESISTANCE MECHANISM

The first-story columns of the test structure were able to maintain their load-carrying capacity over a large deformation range. An envelope of first-story current maximum displacement, attained over the four simulations, is shown in Fig. 6 against the corresponding base shear force. Loading in both directions is lumped into a single locus. The first story of the test specimen yielded during the initial run at a base shear force of approximately 3.4 kips (15.1 kN) and a displacement

of 0.17 in. (4.3 mm), or 1.4% drift ratio. The maximum value of base
shear that the first-story columns were able to develop over the range
of all simulations was 3.9 kips (17.3 kN) at a displacement of 0.38 in.
((.7 mm), which corresponds to a first-story displacement ductility
factor of 2.2 and a drift ratio of 3.1%. Subsequent loading, during one
of the last stable cycles before collapse, generated a maximum
displacement of 0.86 in. (21.8 mm), corresponding to a 7% first-story
drift ratio.

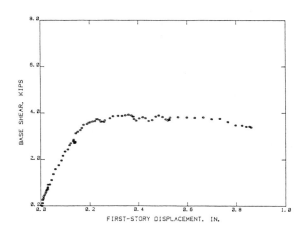

Fig. 6 First-Story Force-Displacement Envelope

The lateral strength of the first story of the test structure was
attained when all first-story columns developed their maximum flexural
resistance. The base shear force at first steel yield, computed assuming
this mechanism, was 13.5 kips (15.6 kN) and agrees well with the observed
quantity of 3.4 kips (15.1 kN). A base shear strength of 3.7 kips (16.5
kN), estimated assuming the column reinforcement achieved its strength,
was 5% lower than the observed maximum value of base shear. It should be
noted that base shear estimates were computed solely on the basis of
flexural capacity of the column cross-sections and did not include any
contributions of velocity-dependent resistance mechanisms. The observed
base shear quantities, on the other hand, were obtained as the total
inertia load applied to the structure and included velocity-dependent
components if these were present in structural response.

The negatively-sloped portions of the first-story force-displacement
envelope (Fig. 6) are due in part to practically nonexistent post-yield
stiffnesses during large-amplitude cycles (Fig. 5). However, this trend
of the response envelope is mostly attributable to column strength
deterioration under large displacement reversals, as can be observed in
the first-story force-displacement relation during the first few cycles

of the "collapse" run (Fig. 5). The force-displacement envelope has a positive post-yield slope for displacements smaller than 0.38 in. (9.7 mm), or ductility factors less than 2.2 and drift ratios below 3.1%. Beyond this deformation level a decrease in the first-story envelope is observed (Fig. 6). The decrease is proportional to displacement and is quite small (approximately 2% of the first-story lateral stiffness at yield) for displacements smaller than 0.75 in. (19 mm), or ductility factors less than 4.5. For larger deformation levels, the decrease in the response envelope is quite severe (17% of the first-story lateral stiffness at yield).

The reader is reminded that the lateral stiffness of a column is affected by the magnitude of the column axial load. First-story column axial loads due to structure weight were 1.28 kips (5.7 kN), assuming equal distribution among all first-story columns. This quantity is one-tenth of the column strength in pure compression, and three-tenths of the axial load capacity at the balanced strain condition. While this axial load is not very large for a column, it is within realistic range of column axial loading for frames in regions of high seismic risk. It should also be noted that the columns of the model structure exhibited very small post-yield flexural stiffnesses due to the mechanical characteristics of the model reinforcement. The strength of the column reinforcement was only 9% larger than the yield stress. This ratio is expected to be at least 25% for typical deformed bars.

DATA PROCESSING

The response of the test structure during the collapse is certainly interesting and definitely surprising. The ability of the first story to develop drift ratios of 7% before collapse would hardly be credible if it were not substantiated. The test results deserve further study in order to gain additional insight on structural response prior to collapse. Specifically, it is of interest to attempt to define the time and displacement at the instant the system became unstable. However, problems with the LVDT signals during the "collapse" run prevent direct study of displacement response. Firstly, the computer channels that recorded the displacement signals were saturated by the large voltages emitted by the displacement transducers after 2.75 sec. Secondly, the first-level displacement transducers were driven out of their linear range before the model structure collapsed.

In order to study structural response just before collapse, it is necessary to reproduce the displacement histories by correcting and integrating the acceleration records. Electrical noise present in the acceleration data produces huge errors in displacement if direct integration is performed on the uncorrected accelerations. It is speculated that the noise arises from two distinct sources. The alternating house current introduced high-frequency components, and offset voltages in the signal conditioning euipment produced low-frequency trends. It is the low-frequency trends that generate large cumulative errors in displacement.

In a study of the earthquake response of snall-scale r/c model structure subjected to base motions, Kreger and Sozen (1983) outlined a method for correcting and integrating acceleration records to produce displacement histories. A band-pass frequency filter was applied to the acceleration data to eliminate low- and high-frequency trends, and a least-squares algorithm was applied to the corrected and integrated acceleration record to identify any linear trends in velocity that had escaped the filter. The data processing scheme was observed to perform quite well in reproducing the motion of the earthquake simulator table, but the accuracy of the method was found to be highly dependent on the bottom cutoff frequency for the band-pass filter window (Kreger and Sozen, 1983).

Toussi and Yao (1983) proposed another scheme for correcting baseline errors in acceleration response histories for structural systems, the polynomial identification method. This technique was later used by Stephens et. al. (1985) in a comparative study of acceleration record correction methods and was found to be as effective as any of the other methods considered. The scheme consists of identifying the baseline error in the uncorrected and integrated acceleration record by least-squares fitting a high-order polynomial to the data. This method not only enables the correction of the velocity data by subtraction of the high-order polynomial, but also yields consistent acceleration histories if the first derivative of the polynomial is subtracted from the uncorrected acceleration record. In their study, Toussi and Yao (1983) used a fifth degree polynomial to identify baseline error. Stephens et. al. (1985) obtained best results from a polynomial of ninth degree.

Both of these schemes were used in the present study to correct and integrate acceleration records, and are referred to as Scheme 1 (Kreger and Sozen, 1983) and Scheme 2 (Toussi and Yao, 1983). Numerical integration was based on Simpson's 1/3 and 3/8 rules (Chapra and Canale, 1985). Frequency filtering of the acceleration waveform was achieved by transforming the record to the frequency domain with a Fast Fourier Transform (FFT) routine, and eliminating the amplitudes of the frequency components above and below the specified cutoff frequencies. The filtered record was transformed back to the time domain with an inverse FFT routine.

Both data processing schemes were applied to the first-run ninth-level acceleration history, after subtraction of the base acceleration record, to compute top displacement relative to the structure base. Because top displacement, relative to the base of the structure, was also measured during the first run, it was used to calibrate and compare the data processing schemes. The best results for Scheme 1 were obtained using bottom and top band-pass filter cutoff frequencies of 0.25 and 25 Hz, respectively. An eighth-order polynomial was found to provide the best performance for Scheme 2.

Top displacement histories for the first run, computed using both schemes, are shown in Fig. 7 along with the measured displacement record. Both schemes effectively eliminated cumulative baseline error while accurately capturing low- and high-frequency components of response. The superior performance of Scheme 2 during the first four sec. of response is clearly evident during the half-cycle of maximum response (2.3 sec.). However, after approximately 4.5 sec. of response the displacement obtained from Scheme 2 has a larger amount of baseline distortion than that obtained from Scheme 1. Both methods appear to be quite effective in handling cumulative baseline error, but in the process they also eliminate residual displacement. While this feature may be desirable for correcting accelerograms, where the ground is assumed to have no residual displacement, it is unacceptable for processing nonlinear structural response records for systems that undergo permanent deformation.

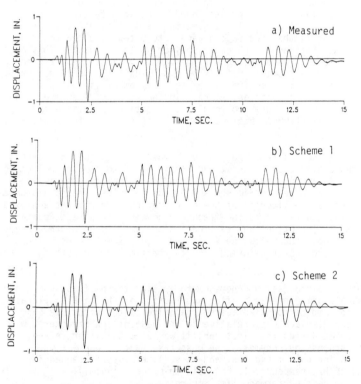

Fig. 7   Computed Top Displacement During Run 1

Nonlinear deformation of the model structure induced a permanent deformation of 0.07 in. (2mm). It can be argued that neither of the two data processing schemes should have eliminated the residual displacement because the filter (Scheme 1) and the baseline correction (both schemes) were applied to the acceleration and velocity histories, respectively.

These two processes eliminate trends in acceleration and velocity, but should not affect a constant shift in displacement. A constant shift in the displacement history should have no derivatives with time, making it transparent to the correction schemes.

Permanent deformation due to nonlinear action in structural systems will not occur as a single-valued quantity over the entire duration of response. The baseline is shifted each time the structure is driven into the nonlinear range, and the magnitude of the shift is proportional to the amount of nonlinear deformation. Thus, for a vibrating nonlinear system baseline shift is a function of time, and as such it can have derivatives with time. It appears that the band-pass filter identifies the baseline shift as a low-frequency component and eliminates it. The polynomial identification scheme also acts as a frequency filter, due to the wavelike nature of polynomials, and also appears to eliminate the baseline shift that was induced by permanent deformation.

A correction scheme that discriminates between baseline error and baseline shifts due to permanent deformations is needed to reproduce displacement response during the "collapse" run. It is not essential that the scheme guarantee a high degree of accuracy over an extended duration, but it is crucial that it preserve permanent deformation. In order to identify the true baseline error, the following procedure is proposed. Double integration of the uncorrected acceleration histories of the model structure, relative to the base, produce a displacement record that includes the cumulative baseline error. Subtraction of the measured displacement record for that story yields a displacement error function which does not include the baseline shift due to nonlinear deformation. The first and second derivatives of the displacement error function give the velocity and acceleration error functions. This error identification procedure can only be applied to those earthquake simulations for which displacement records have been measured.

If a consistent and quantifiable trend can be established in the displacement, velocity and acceleration error functions, an analytical curve will be fitted to the data. Extrapolation of this curve and its derivatives to the "collapse" run will enable the calculation of the displacement records. If no consistent trend can be identified or quantified in the error functions for the first three runs, then the displacement data that is available for the "collapse" run will be used to define error functions for the first 2.75 sec. of response. Extrapolation of these error functions for an additional 0.55 sec. should serve to provide corrected displacement data until collapse.

SUMMARY AND CONCLUSIONS

The results of simulated earthquake tests of a small-scale model of a nine-story r/c frame structure were presented. The response of the test specimen was characterized by inelastic deformations of the first-story columns and large first-story drifts. During the fourth simulated earthquake, which had a nominal intensity four times as large as the "design" motion, the test structure exhibited a maximum first-story drift ratio of 7% before collapse. Excessive lateral deformation of the first story resulted in an instability failure. Due to limitations of the instrumentation, recording of displacements ceased approximately 0.55 sec. before collapse.

The first-story columns of the model structure were able to maintain their load-carrying capacity mostly intact over a large deformation range ($\leq$ 6% drift ratio, or 4.5 ductility factor) under reversing loads. Marked strength deterioration under reversing loads was observed for displacement ductility factors larger than 4.5. The tests certainly confirm the obvious fact that SBWC frames are susceptible to instability due to large lateral deflections and column strength deterioration under large displacement reversals. However, the tests also serve to demonstrate that for some frame structures of the SBWC type, there can exist a large range of base motion intensity separating adequate behavior from intolerable response.

Two different schemes were used to reproduce displacement histories from measured acceleration records. Besides elimination of the permanent deformation due to nonlinear action, the schemes were very effective in eliminating baseline error. Because it is known that extensive nonlinear deformation was induced in the first-story columns o the model structure just before collapse, it was decided that neither of the two schemes was adequate for producing displacement records. A procedure was proposed which promises to provide a correction technique that does not eliminate permanent deformation.

If the displacement response of the test structure can be defined up to collapse, further study can be carried out. Because static stability differs from stability in the dynamic sense, proper consideration of dynamic response is required to establish the onset of instability during the "collapse" run. It may be possible to identify the time and displacement at the instant of instability by concurrent inspection of the first-story velocity and displacement histories.

REFERENCES

1. Chapra, S.C., and Canale, R.P., Numerical Methods for Engineers With Personal Computer Applications, McGraw-Hill, Inc., N.Y., N.Y., 1985.

2. Housner, G.W., "Behavior of Structures During Earthquakes," Journal of the Engineering Mechanics Division, ASCE, Vol. 85, No. EM4, October 1959, pp. 108-129.

3. Kreger, M.E., and Sozen, M.A., "A Study of the Causes of Column Failures in the Imperial County Services Building During the 15 October 1979 Imperial Valley Earthquake," Civil Engineering Studies. Structural Research Series No. 509, University of Illinois, Urbana, August 1983.

4. Schultz, A.E. "An Experimental and Analytical Study of the Earthqua Response of R/C Frames with Yielding Columns," Ph.D. Dissertation submitted to the Graduate College of the University of Illinois, Urbana, May 1986.

5. Shibata, A., and Sozen, M.A., "Substitute-Structure Method for
    Seismic Design in R/C," _Journal of the Structural Division_, ASCE,
    Vol. 102, No. ST1, January 1976, pp. 1-18.

6. Sozen, M.A., Otani, S., Gulkan, P., and Nielsen, N.M., "The
    University of Illinois Earthquake Simulator," _Proceedings_, Fourth
    World Conference on Earthquake Engineering, Santiago, Chile, January
    1969, Vol. III, Session B5.

7. Stephens, J.E., Brady, G.A., and Yao, J.T.P., "Data Processing in
    Earthquake Engineering," Structural Engineering Report No. CE-STR-
    85-40, Purdue University, West Lafayette, Indiana, 1985.

8. Toussi, S., and Yao, J.T.P., "Hysteresis Identification of Existing
    Structures," _Journal of the Engineering Mechanics Division_, ASCE,
    Vol. 109, No. EM5, October 1983, pp. 1189-1202.

SUBJECT INDEX

Page number refers to first page of paper.

AUTHOR INDEX

Page number refers to first page of paper.